W9-CAJ-380

STUDENT WORKBOOK FOR

Modeling, Functions, and Graphs THIRD EDITION

ALGEBRA FOR COLLEGE STUDENTS

Katherine Yoshiwara
Bruce Yoshiwara

Los Angeles Pierce College

BROOKS/COLE

™

THOMSON LEARNING

Australia • Canada • Mexico • Singapore • Spain • United Kingdom • United States

BROOKS/COLE

THOMSON LEARNING

Assistant Editor: *Rachael Sturgeon*
Marketing Team: *Leah Thomson, Maria Salinas*
Production Coordinator: *Dorothy Bell*

Cover Design: *Roy R. Neuhaus*
Print Buyer: *Micky Lawler*
Printing and Binding: *Globus Printing*

For more information about this or any other Brooks/Cole product, contact:
BROOKS/COLE
511 Forest Lodge Road
Pacific Grove, CA 93950 USA
www.brookscole.com
1-800-423-0563 (Thomson Learning Academic Resource Center)

Printed in the United States of America

10 9 8 7 6 5 4 3 2

ISBN 0-534-39631-3

Note to Students

This workbook contains tables and grids for the activities and problems in your textbook. The grids are already labeled with appropriate scales for the graphs you will draw. We hope that providing these grids will eliminate some of the time-consuming work involved in drawing a graph, and allow you to concentrate on the mathematics.

The lessons in your textbook include Exercises printed on a blue background. There are copies of these Exercises in the workbook, with space for you to show your work and record your answers. You should try these Exercises as you read the text, to see if you understand the material.

You should try to keep your workbook up to date as your course progresses. You will also need a spiral or loose-leaf notebook for class notes and the rest of your homework problems. Your workbook and notebook will be useful study aids when you are preparing for exams.

How to Be Successful in Your Math Class

The key to success in a math class (as in most endeavors) is persistence. You cannot learn mathematics in one great rush the night before the exam; but you can master it in small chunks a little at a time. You should plan to study math for at least one hour every night. Don't give up until you have a good grasp of the lesson and can work the problems on your own. If you get behind in a math class it is very difficult to catch up.

1. **Attend class every day.**
   Studies have shown that success in math classes is correlated strongly with attendance. If you must miss a class, find out beforehand what the class will cover. Read the lesson and complete the assignment anyway, just as if you had attended.
   a. Use class time wisely. This is your best opportunity to learn the material.
   b. Take notes. Learn to summarize what the instructor says, not just what he or she writes on the board.
   c. Don't be afraid to ask questions when you don't follow the lesson.

2. **Read the text book.**
   Reading a math book is not like reading a novel. You will need to read the material more than once to understand and retain it.
   a. Read the new material *before* it will be covered in class.
   b. Read with a pencil in hand so that you can make notes to yourself, underline important points, or put question marks in the margins.
   c. Read the section again after it has been covered in class.

3. **Look over your handouts and class notes.**

The sooner you can review your notes after class, the better. People forget most of what they hear very quickly, and reviewing your notes will help you retain the new information.

   a. Look for points where your notes reinforce the material in the textbook.

   b. Try to fill in any steps or information you may have missed in class.

   c. Write a sentence or two summarizing the main points of the lesson.

4. **Do the homework problems.**

Most of your learning takes place when you work problems. If you do some of your work in a study group or tutoring center, you will have someone to consult as soon as you hit a snag.

   a. If you get stuck on a problem, refer to the textbook or your notes for help.

   b. Call a classmate on the phone and try to figure out together the problems you had trouble with.

   c. Mark any problems you can't get, but don't stop! Skip those problems for now, and continue on to the end of the assignment.

5. **Get help right away.**

Mathematics builds upon earlier material, so if you don't understand today's lesson you will have even more trouble tomorrow or the next day.

   a. Make a list of points you don't understand and problems you need help with.

   b. Ask your instructor or a tutor for help *today* -- don't put it off!

   c. Fill in your notes with the answers to your questions, and make sure you can work all the problems that gave you trouble.

6. **Prepare for exams.**

In addition to keeping up with daily work, you must prepare specifically for exams. Always study 100% of the material the exam will cover. If you omit some topics, you won't be sure which problems you should work on during the exam!

   a. Begin studying for the exam a week ahead of time, so that you will have a chance to get help on any topics you are unsure about.

   b. Make a check-list or outline of the material the exam will cover, and review each topic until you have mastered it.

   c. Have a classmate or tutor make up a sample exam (or make one yourself), and practice working problems under exam conditions.

Table of Contents

Chapter 1   Linear Models

Investigation 1 Sales on Commission

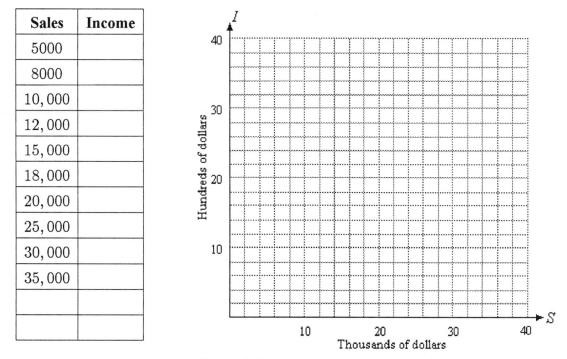

Sales	Income
5000	
8000	
10,000	
12,000	
15,000	
18,000	
20,000	
25,000	
30,000	
35,000	

Table 1.1

Figure 1.1

Section 1.1 Some Examples of Linear Models

Exercise 1   Frank plants a dozen corn seedlings, each 6 inches tall. With plenty of water and sunlight they will grow approximately 2 inches per day. Complete the table of values for the height, h, of the seedlings after t days.

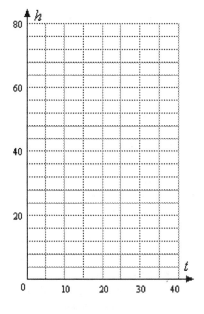

t	0	5	10	15	20
h					

a. Write an equation for the height h of the seedlings in terms of the number of days t since they were planted.

b. Graph the equation on the grid.
c. How tall is the corn after 3 weeks?
d. How long will it be before the corn is 6 feet tall?

1

Exercise 2 Silver Lake has been polluted by industrial waste products. The concentration of toxic chemicals in the water is currently 285 parts per million (ppm). Local environmental officials would like to reduce the concentration by 15 ppm each year.

    a. Complete the table of values showing the concentration, C, of toxic chemicals t years from now. For each t-value, calculate the corresponding value for C. Write your answers as ordered pairs.

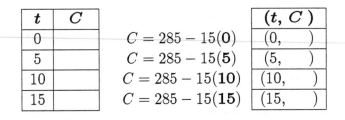

t	C			(t, C)
0		$C = 285 - 15(0)$		(0,)
5		$C = 285 - 15(5)$		(5,)
10		$C = 285 - 15(10)$		(10,)
15		$C = 285 - 15(15)$		(15,)

    b. Graph the ordered pairs on the grid, and connect them with a straight line. Extend the graph until it reaches the horizontal axis, but no farther. Points with negative C-coordinates have no meaning for the problem.

    c. Write an equation for the concentration, C, of toxic chemicals t years from now. The concentration is initially 285 ppm, and we *subtract* 15 ppm for each year that passes, or $15 \times t$.

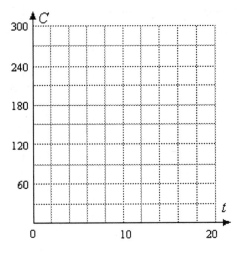

Exercise 3 Find the intercepts of the graph of
$$2y = -18 - 3x.$$

To find the y-intercept, set $x = 0$ and solve for y:

To find the x-intercept, set $y = 0$ and solve for x:

x	y
0	
	0

2

Exercise 4a. Find the intercepts of the graph of

$$60x - 13y = 390.$$

x	y
0	
	0

b. Use the intercepts to help you choose appropriate scales for the axes, and graph the equation.

Homework 1.1

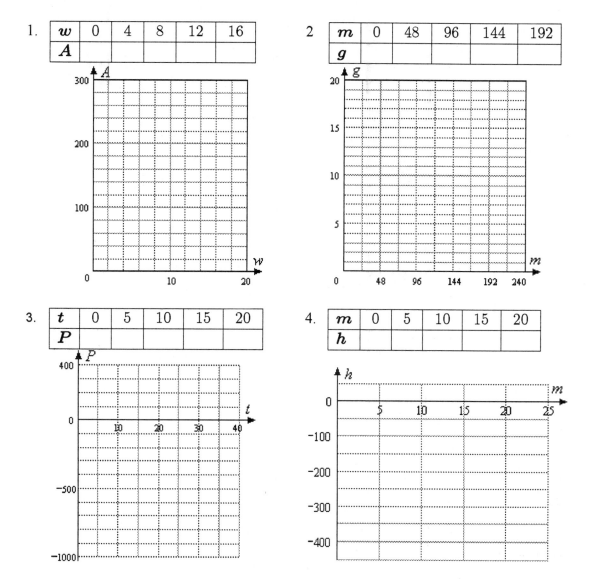

1.

w	0	4	8	12	16
A					

2

m	0	48	96	144	192
g					

3.

t	0	5	10	15	20
P					

4.

m	0	5	10	15	20
h					

5. $x + 2y = 8$

6. $2x - y = 6$

7. $3x - 4y = 12$

8. $2x + 6y = 6$

9. $\dfrac{x}{9} - \dfrac{y}{4} = 1$

10. $\dfrac{x}{5} + \dfrac{y}{8} = 1$

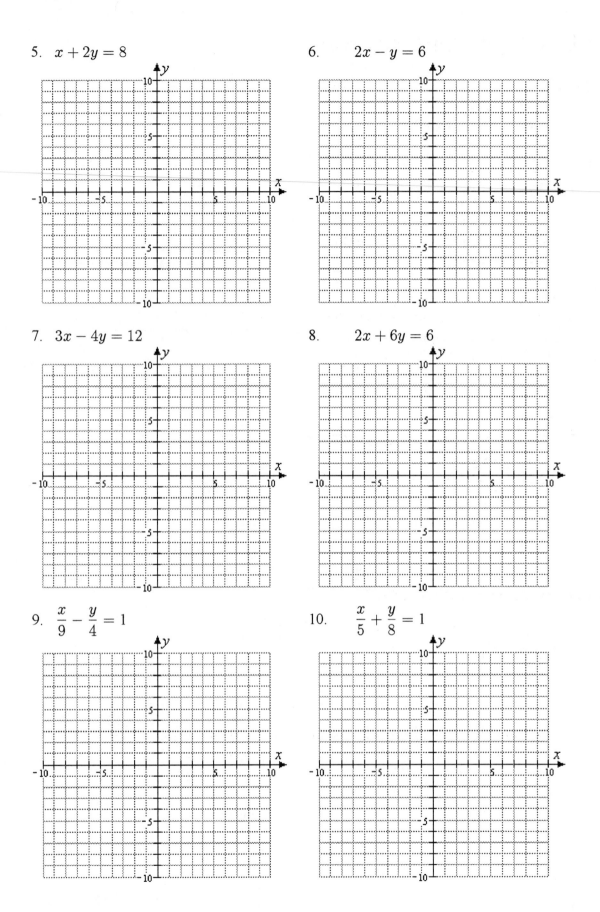

1.2 Using a Graphing Calculator

Exercise 1a. Solve the equation $7 - 2y = 4x$ for y in terms of x.

Subtract 7 from both sides :

Divide both sides by -2 :

Simplify :

b. Graph the equation in the standard window.
Press $\boxed{Y=}$ and enter the equation.
Then press $\boxed{\text{ZOOM}}$ $\boxed{6}$.

Exercise 2a. Find the x- and y-intercepts of the graph of $2y - 1440 = 45x$.

x	y
0	
	0

b. Graph the equation on a graphing calculator. Choose a window that shows both of the intercepts.

Xmin =   Xmax =

Ymin =   Ymax =

Exercise 3a. Graph the equation

$$y = 32x - 42$$

in the window

Xmin $= -4.7$ Xmax $= 4.7$ Xscl $= 1$
Ymin $= -250$ Ymax $= 50$ Yscl $= 1$

Use the $\boxed{\text{TRACE}}$ feature to find the point that has y-coordinate -122.

b. Verify your answer algebraically by substituting your x-value into the equation.

Exercise 4a. Use the graph of $y = 30 - 8x$ to solve the equation $30 - 8x = 50$.

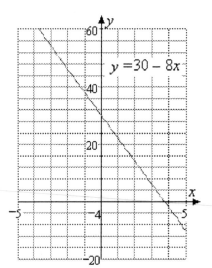

    b. Verify your solution algebraically.

Exercise 5a. Use the graph of $y = 30 - 8x$ to solve the inequality $30 - 8x < 14$.

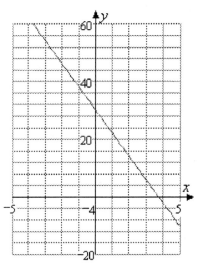

    b. Verify your solution algebraically.

Exercise 6a. Graph the equation $y = 1.3x + 2.4$. Set the WINDOW to

$$\text{Xmin} = -4.6 \quad \text{Xmax} = 4.8,$$
$$\text{Ymin} = -10 \quad \text{Ymax} = 10.$$

    b. Use your graph to solve the inequality $1.3x + 2.4 < 8.51$.

Homework 1.2

17. Figure 1.17 shows a graph of $y = -2x + 6$.
    a. Use the graph to find all values of x for which:
        i) $y = 12$ ii) $y > 12$ iii) $y < 12$
    b. Use the graph to solve:
        i) $-2x + 6 = 12$ ii) $-2x + 6 > 12$ iii) $-2x + 6 < 12$
    c. Explain why your answers to parts (a) and (b) are the same.

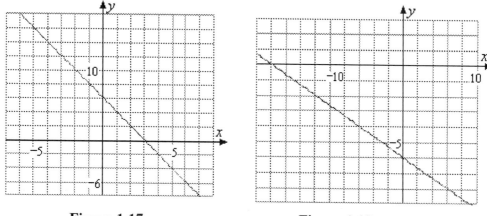

Figure 1.17   **Figure 1.18**

18. Figure 1.18 shows a graph of $y = \dfrac{-x}{3} - 6$.
    a. Use the graph to find all values of x for which:
       i) $y = -4$ ii) $y > -4$ iii) $y < -4$
    b. Use the graph to solve:
       i) $\dfrac{-x}{3} - 6 = -4$ ii) $\dfrac{-x}{3} - 6 > -4$ iii) $\dfrac{-x}{3} - 6 < -4$
    c. Explain why your answers to parts (a) and (b) are the same.

19. Figure 1.19 shows the graph of $y = 1.4x - 0.64$. Solve:
    a. $1.4x - 0.64 = 0.2$ b. $-1.2 = 1.4x - 0.64$
    c. $1.4x - 0.64 > 0.2$ d. $-1.2 > 1.4x - 0.64$

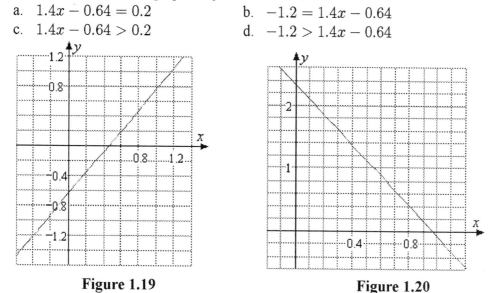

Figure 1.19   **Figure 1.20**

20. Figure 1.20 shows the graph of $y = -2.4x + 2.32$. Solve:
    a. $1.6 = -2.4x + 2.32$ b. $-2.4x + 2.32 = 0.4$
    c. $-2.4x + 2.32 \geq 1.6$ d. $0.4 \geq -2.4x + 2.32$

1.3 Slope

Exercise 1   Compute the slope of the line through the indicated points. On both axes, one square represents one unit.

$$\frac{\text{change in } y\text{-coordinate}}{\text{change in } x\text{-coordinate}} =$$

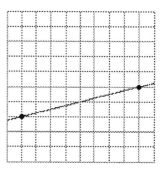

Exercise 2a.   Graph the line $4x - 2y = 8$ by finding the x- and y-intercepts.

x	y
0	
	0

b.   Compute the slope of the line using the x-intercept and y-intercept. Move from $(0, -4)$ to $(2, 0)$ along the line.

$$m = \frac{\Delta y}{\Delta y} =$$

c.   Compute the slope of the line using the points $(4, 4)$ and $(1, -2)$.

$$m = \frac{\Delta y}{\Delta y} =$$

Exercise 3   The graph shows the altitude a (in feet) of a skier t minutes after getting on a ski lift.

a.   Choose two points and compute the slope (including units).

$$m = \frac{\Delta a}{\Delta t} =$$

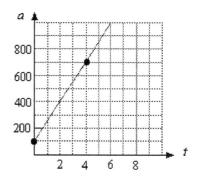

b.   Explain what the slope measures in the context of the problem.

9

Homework 1.3

9.

10.

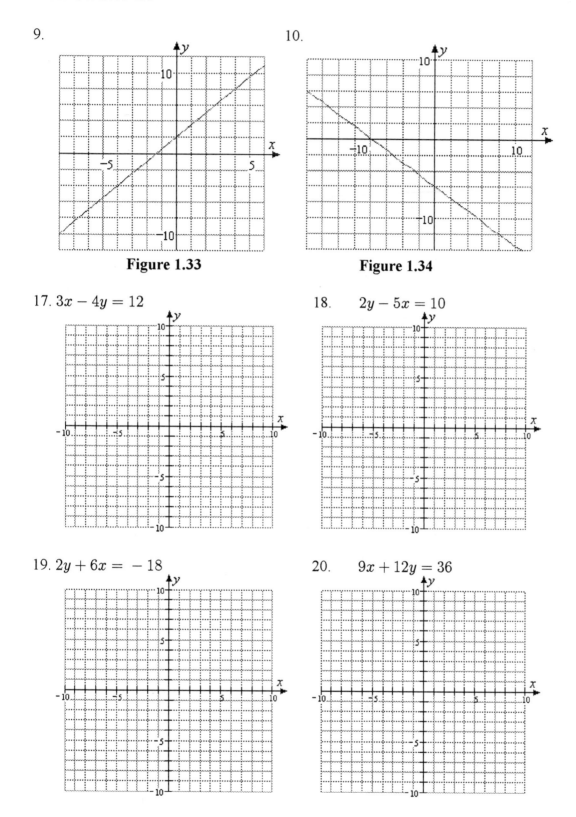

Figure 1.33

Figure 1.34

17. $3x - 4y = 12$

18. $2y - 5x = 10$

19. $2y + 6x = -18$

20. $9x + 12y = 36$

21. $\dfrac{x}{5} - \dfrac{y}{8} = 1$

22. $\dfrac{x}{7} - \dfrac{y}{4} = 1$

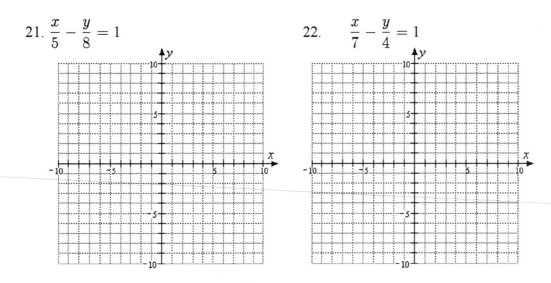

27. A temporary typist's paycheck (before deductions) in dollars is given by $S = 8t$, where t is the number of hours she worked.

    a. Make a table of values for the equation.

t	4	8	20	40
S				

    b. Graph the equation.
    c. Using two points on the graph, compute the slope $\dfrac{\Delta S}{\Delta t}$, including units.
    d. What is the significance of the slope in terms of the typist's paycheck?

28. The distance in miles covered by a cross-country competitor is given by $d = 6t$, where t is the number of hours she runs.

    a. Make a table of values for the equation.

t	2	4	6	8
d				

    b. Graph the equation.
    c. Using two points on the graph, compute the slope $\dfrac{\Delta d}{\Delta t}$, including units.
    d. What is the significance of the slope in terms of the cross-country runner?

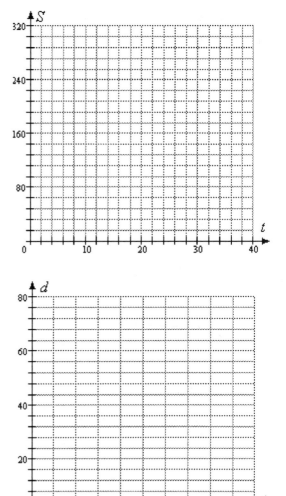

Midchapter Review

1. In the desert the temperature at 6 a.m., just before sunrise, was 65° F. The temperature rose 5 degrees every hour until it reached its maximum value at about 5 p.m. Complete the table of values for the temperature T at h hours after 6 a.m.

h	0	3	6	9	10
T					

   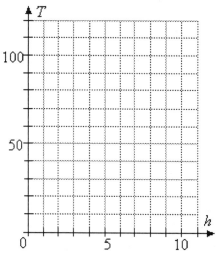

   a. Write an equation for the temperature T in terms of h.

   b. Graph the equation.
   c. How hot is it at noon?

   d. When will the temperature be 110°F?

2. The taxi out of Dulles Airport charges a traveler with one suitcase an initial fee of $2.00, plus $1.50 for each mile traveled. Complete the table of values showing the charge, C, for a trip of n miles.

n	0	5	10	15	20	25
C						

   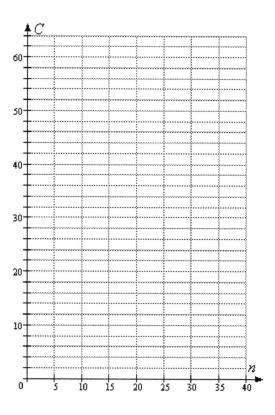

   a. Write an equation for the charge C in terms of the number of miles traveled, n.

   b. Graph the equation.
   c. What is the charge for a trip to Mount Vernon, 40 miles from the airport?

   d. If a ride to the National Institute of Health costs $39.50, how far is it from the airport to the NIH?

13

3. Delbert's altitude is 300 meters when he begins to descend in his hot air balloon. He descends by 5 meters each minute.

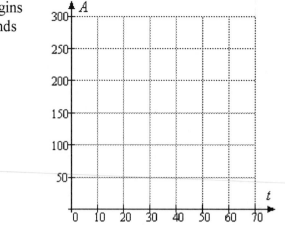

t	A
0	
10	
20	
30	
40	

a. Complete the table of values showing Delbert's altitude, A, t minutes after starting his descent. Graph the data points and connect them with a straight line. Extend the graph until it reaches the horizontal axis.

b. Write an equation for Delbert's altitude, A, in terms of t.

c. Explain the significance of the intercepts to the problem situation.

4. Francine received 120 hours of free Internet connect time as an introductory membership offer from Yippee.com. She spends 1.5 hours per day connected to the Internet.

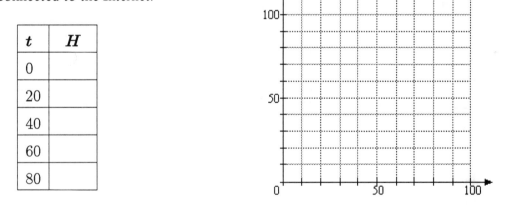

t	H
0	
20	
40	
60	
80	

a. Complete the table of values showing the number of free hours, H, that Francine has left t days after starting her membership. Graph the data points and connect them with a straight line. Extend the graph until it reaches the horizontal axis.

b. Write an equation for the remaining free hours, H, in terms of t.

c. Explain the significance of the intercepts for the problem situation.

14

9. a. Use the graph in Figure 1.45 to solve $0.24x - 3.44 = -2$.
    b. Verify your answer algebraically.
    c. Use the graph to solve $0.24x - 3.44 > -2$.
    d. Solve the inequality in part (c) algebraically.

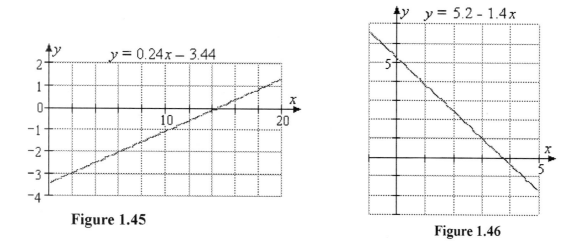

Figure 1.45

Figure 1.46

10. a. Use the graph in Figure 1.46 to solve $5.2 - 1.4x = 1$.
    b. Verify your answer algebraically.
    c. Use the graph to solve $5.2 - 1.4x > 1$.
    d. Solve the inequality in part (c) algebraically.

1.4   Equations of Lines

Exercise 1a. Write the equation

$$2y + 3x + 4 = 0$$

in slope-intercept form.

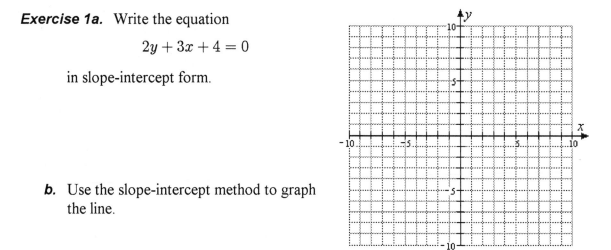

b. Use the slope-intercept method to graph the line.

Exercise 2a. Find the slope of the line passing through the points $(2, -3)$ and $(-2, -1)$.

$m =$

b. Make a rough sketch of the line by hand.

Exercise 3   Find an equation for the line shown.

$b =$

$m =$

$y =$

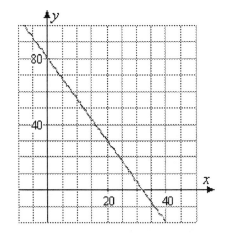

Exercise 4   Use the point-slope form to find the equation of the line that passes through the point $(-3, 5)$ and has slope -1.4.

$$y - y_1 = m(x - x_1)$$ Substitute -1.4 for m and $(-3, 5)$ for (x_1, y_1).

Simplify: apply the distributive law.

Solve for y.

Homework 1.4

11. $m = 3$ and $b = -2$ 12. $m = -4$ and $b = 1$

13. $m = -\dfrac{5}{3}$ and $b = -6$ 14. $m = \dfrac{3}{4}$ and $b = -2$

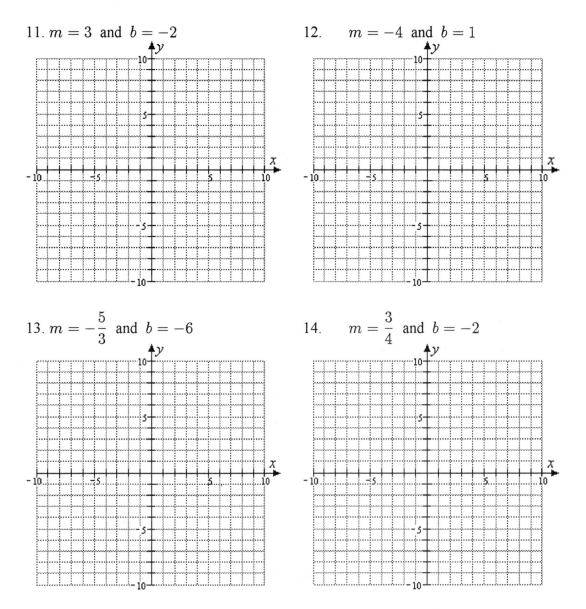

23. In England, oven cooking temperatures are often given as Gas Marks rather than degrees Fahrenheit. The table shows the equivalent oven temperatures for various Gas Marks.

Gas Mark	3	5	7	9
Degrees (F)	325	375	425	475

a. Plot the data and draw a line through the data points.
b. Calculate the slope of your line. Estimate the y-intercept from the graph.

c. Find an equation that gives the temperature in degrees Fahrenheit in terms of the Gas Mark.

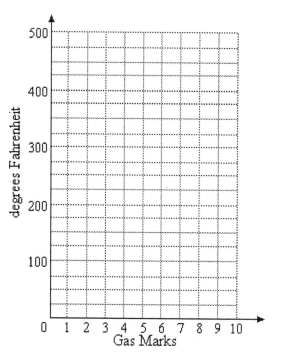

24. European shoe sizes are scaled differently than American shoe sizes. The table shows the European equivalents for various American shoe sizes.

American shoe size	5.5	6.5	7.5	8.5
European shoe size	37	38	39	40

a. Plot the data and draw a line through the data points.

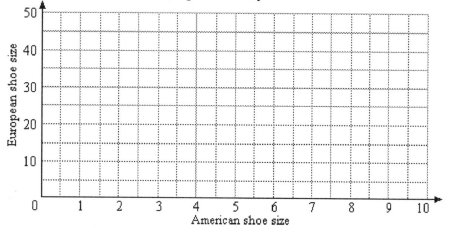

b. Calculate the slope of your line. Estimate the y-intercept from the graph.

c. Find an equation that gives the European shoe size in terms of American shoe size.

19

33. $(2, -5)$; $m = -3$

34. $(-6, -1)$; $m = 4$

35. $(2, -1)$; $m = \frac{5}{3}$

36. $(-1, 2)$; $m = -\frac{3}{2}$

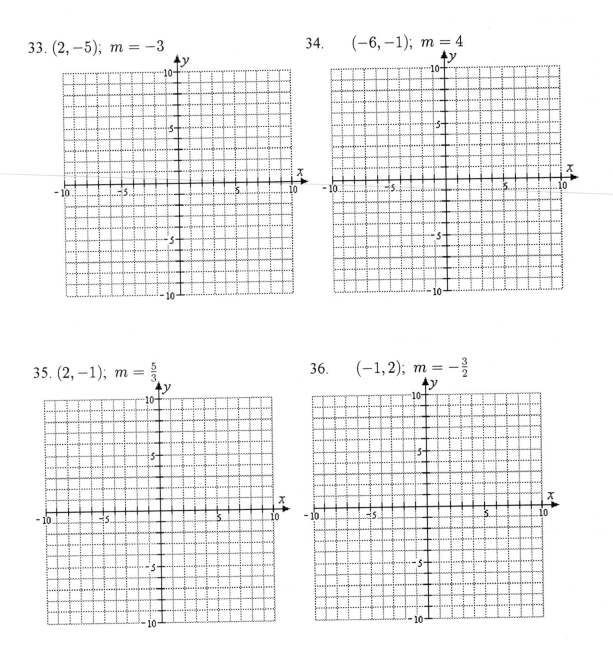

1.5   Lines of Best Fit

Exercise 1   High-frequency radiation is harmful to living things because it can cause changes in their genetic material. The data below, collected by C. P. Oliver in 1930, show the frequency of genetic transmutations induced in fruit flies by doses of x-rays, measured in roentgens. [C.P. Oliver, 1930]

Dosage (roentgens)	285	570	1640	3280	6560
Percent of mutated genes	1.18	2.99	4.56	9.63	15.85

a.   Plot the data on the grid.

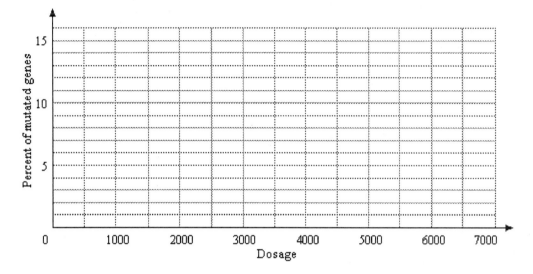

b.   Draw a line of best fit through the data points.
c.   Find the equation of your regression line.

d.   Use the regression equation to predict the percent of mutations that might result from exposure to 5000 roentgens of radiation.

21

Exercise 2a. The temperature in Delbert's apartment was 35°C at 1 p.m. when he turned on the air conditioning, and by 5 p.m. it had dropped to 29°C . Find a linear equation that approximates the temperature h hours after noon

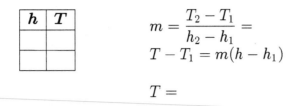

h	T

$$m = \frac{T_2 - T_1}{h_2 - h_1} =$$

$$T - T_1 = m(h - h_1)$$

$$T =$$

b. Use linear interpolation to estimate the temperature at 2 p.m., and extrapolate to predict the temperature at 8 p.m.

Exercise 3a. Use your calculator's statistics features to find the least squares regression equation for the data in Exercise 1.
b. Plot the data and the graph of the regression equation.
c. Use the regression equation to predict the number of mutations expected from a 5000 roentgen dose of radiation.

Homework 1.5

1.   3.

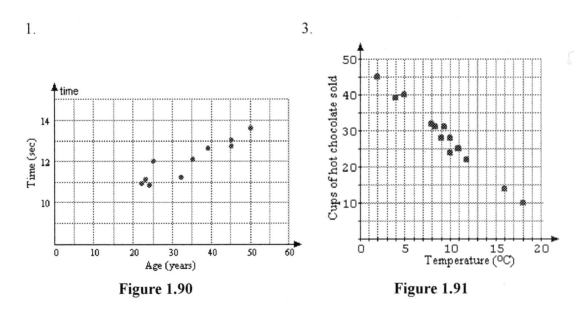

Figure 1.90   **Figure 1.91**

3.

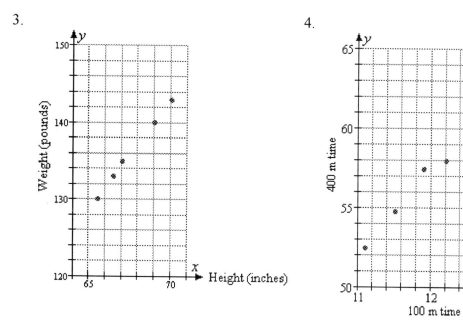

Figure 1.92

4.

Figure 1.93

8.

9.

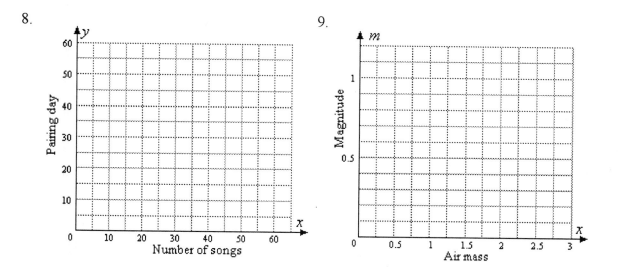

10.

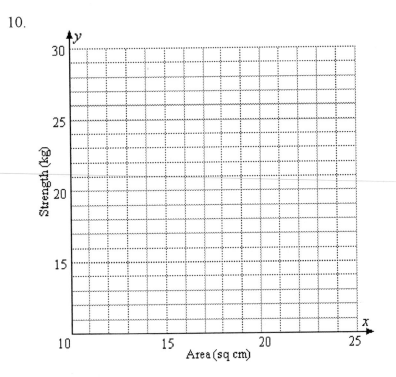

11.

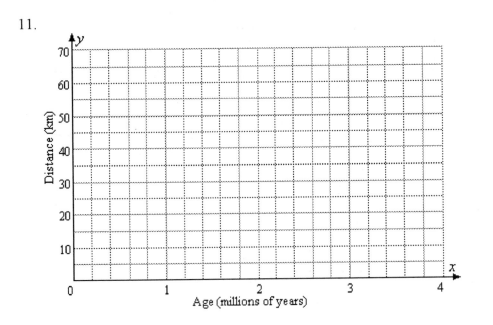

12.

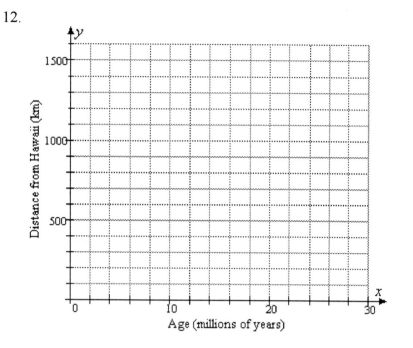

13.

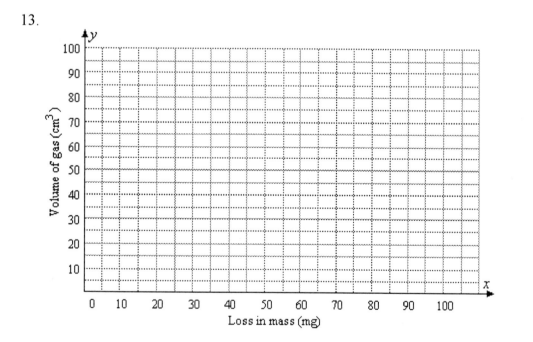

14.

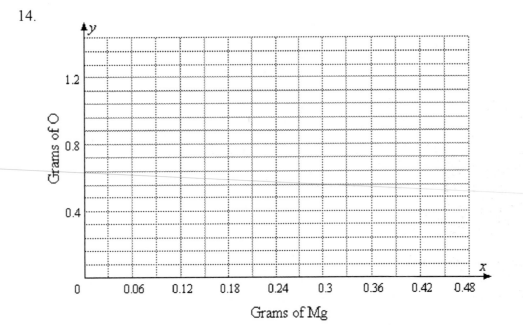

15.

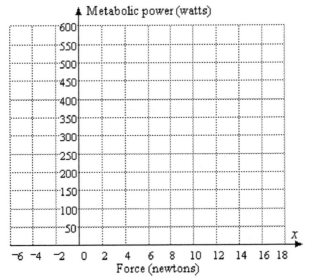

16.

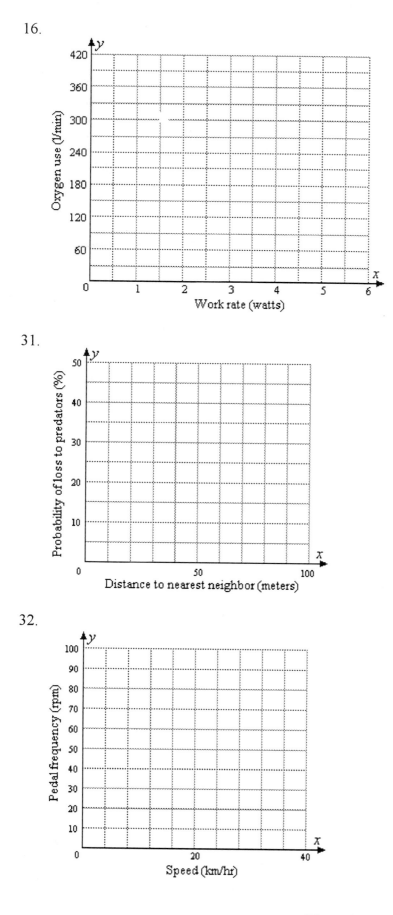

31.

32.

1.6   Additional Properties of Lines

Exercise 1a. What is the slope of any line parallel to the x-axis?

b. What is the slope of any line parallel to the y-axis?

Exercise 2   Are the graphs of $3x - 5y = 5$ and $2y = \dfrac{10}{3} x + 3$ perpendicular?

Put each equation into slope-intercept form by solving for y.

$$3x - 5y = 5 \qquad\qquad\qquad 2y = \frac{10}{3} x + 3$$

Read the slope of each line from its equation.

$$m_1 = \qquad\qquad\qquad\qquad m_2 =$$

Compute the negative reciprocal of m_1: $\dfrac{1}{m_1} =$

Does $m_2 = \frac{-1}{m_1}$?

Exercise 3   Find an equation for the altitude of the triangle shown.

The altitude of a triangle is perpendicular to its base.
Find the slope of the base.

$$m_1 =$$

Find the slope of the altitude.

$$m_2 =$$

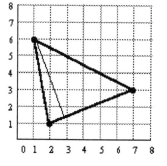

Use the point-slope formula to find the equation of the altitude. Use m_2 for the slope,
and the vertex of the triangle for (x_1, y_1).
$$y - y_1 = m(x - x_1)$$

29

21. a. Sketch the triangle with vertices $A(2,5)$, $B(5,2)$, and $C(10,7)$.
    b. Show that the triangle is a right triangle. (*Hint*: What should be true about the slopes of the two sides that form the right angle?)

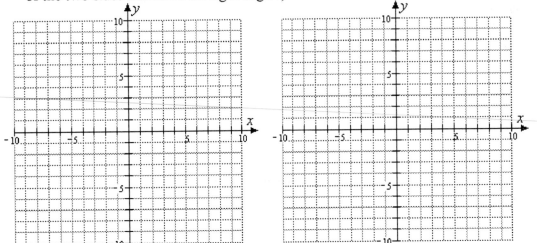

22. a. Sketch the triangle with vertices $P(-1,3)$, $Q(-3,8)$, and $R(4,5)$.
    b. Show that the triangle is a right triangle. (See the hint for Problem #21.)

23. a. Sketch the quadrilateral with vertices $P(2,4)$, $Q(3,8)$, $R(5,1)$ and $S(4,-3)$.
    b. Show that the quadrilateral is a parallelogram. (*Hint*: What should be true about the slopes of the opposite sides of the parallelogram?)

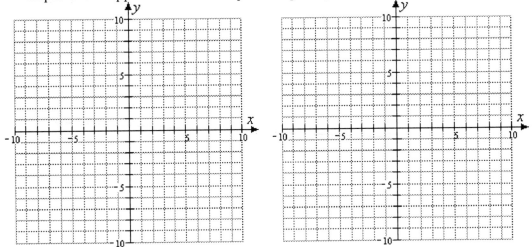

24. a. Sketch the quadrilateral with vertices $A(-5,4)$, $B(7,-11)$, $C(12,25)$ and $D(0,40)$.
    b. Show that the quadrilateral is a parallelogram. (See the hint for Problem #23.)

27. a. Put the equation $x - 2y = 5$ into slope-intercept form, and graph the equation.
    b. What is the slope of any line that is parallel to $x - 2y = 5$?
    c. On your graph for part (a), sketch by hand a line that is parallel to $x - 2y = 5$ and passes through the point $(2, -1)$.
    d. Use the point-slope formula to write an equation for the line that is parallel to the graph of $x - 2y = 5$ and passes through the point $(2, -1)$.

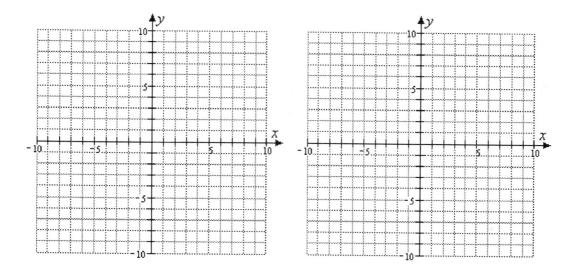

28. a. Put the equation $2y - 3x = 5$ into slope-intercept form, and graph the equation.
    b. What is the slope of any line that is parallel to $2y - 3x = 5$?
    c. On your graph for part (a), sketch by hand a line that is parallel to $2y - 3x = 5$ and passes through the point $(-3, 2)$.
    d. Use the point-slope formula to write an equation for the line that is parallel to the graph of $2y - 3x = 5$ and passes through the point $(-3, 2)$.

31

29. a. Put the equation $2y - 3x = 5$ into slope-intercept form, and graph the equation.
   b. What is the slope of any line that is perpendicular to $2y - 3x = 5$?
   c. On your graph for part (a), sketch by hand a line that is perpendicular to $2y - 3x = 5$ and passes through the point $(1, 4)$.
   d. Use the point-slope formula to write an equation for the line that is perpendicular to the graph of $2y - 3x = 5$ and passes through the point $(1, 4)$.

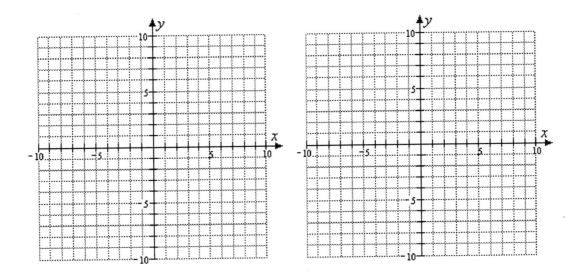

30. a. Put the equation $x - 2y = 5$ into slope-intercept form, and graph the equation.
   b. What is the slope of any line that is perpendicular to $x - 2y = 5$?
   c. On your graph for part (a), sketch by hand a line that is perpendicular to $x - 2y = 5$ and passes through the point $(4, -3)$.
   d. Use the point-slope formula to write an equation for the line that is perpendicular to the graph of $x - 2y = 5$ and passes through the point $(4, -3)$.

33. a. Sketch the triangle with vertices $A(-6,-3)$, $B(-6,3)$, and $C(4,5)$.
    b. Find the slope of the side $\overline{AC}$.
    c. Find the slope of the altitude from point B to side $\overline{AC}$.
    d. Find an equation for the line that includes the altitude from point B to side $\overline{AC}$.

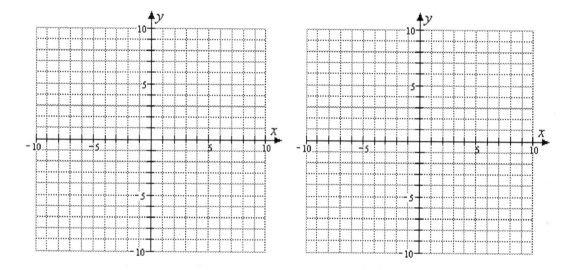

34. a. Sketch the triangle with vertices $A(-5,12)$, $B(4,-2)$, and $C(1,-6)$.
    b. Find the slope of the side $\overline{AC}$.
    c. Find the slope of the altitude from point B to side $\overline{AC}$.
    d. Find an equation for the line that includes the altitude from point B to side $\overline{AC}$.

Chapter 1 Review

1. Last year Pinwheel Industries introduced a new model calculator. It cost $2000 to develop the calculator and $20 to manufacture each one.
    a. Complete the table of values showing the total cost, C, of producing n calculators.

n	100	500	800	1200	1500
C					

    b. Write an equation that expresses C in terms of n.

    c. Graph the equation by hand.
    d. What is the cost of producing 1000 calculators? Illustrate this as a point on your graph.

    e. How many calculators can be produced for $10,000? Illustrate this as a point on your graph.

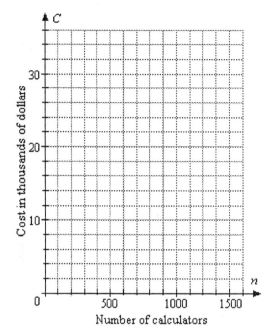

2. Megan weighed 5 pounds at birth and gained 18 ounces per month during her first year.
    a. Complete the table of values that shows Megan's weight, w, in terms of her age, m, in months.

m	2	4	6	9	12
w					

    b. Write an equation that expresses w in terms of m.

    c. Graph the equation by hand.
    d. How much did Megan weigh at 9 months? Illustrate this as a point on your graph.

    e. When did Megan weigh 9 pounds? Illustrate this as a point on your graph.

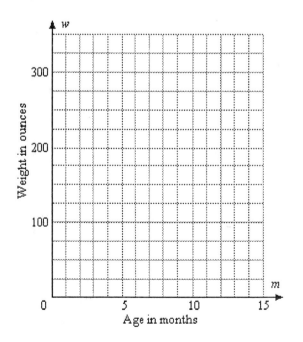

35

59.

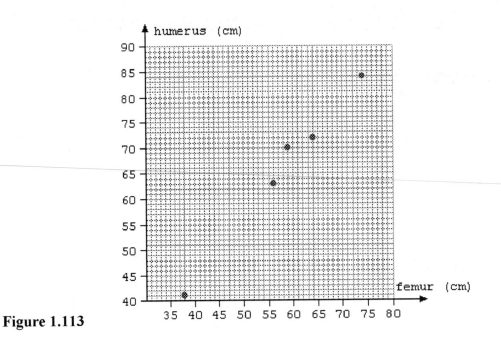

Figure 1.113

60.

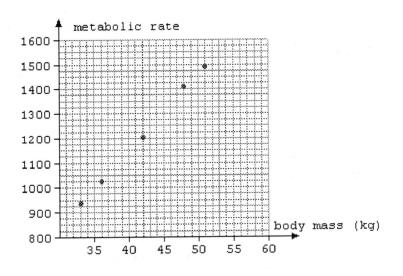

Figure 1.114

Chapter 2   Applications of Linear Models

Investigation 3 Water Level

t (minutes)	Lower Lock Water Level	Upper Lock Water Level
0		
2		
4		
6		
8		
10		

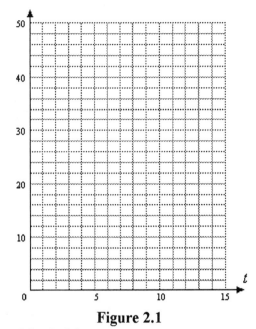

Figure 2.1

2.1   Systems of Linear Equations in Two Variables

Exercise 1a. Solve the system of equations

$$y = -0.7x + 6.9$$
$$y = 1.2x - 6.4$$

by graphing. Use the "friendly" window

$$\text{Xmin} = -9.4 \quad \text{Xmax} = 9.4$$
$$\text{Ymin} = -10 \quad \text{Ymax} = 10$$

b. Verify algebraically that your solution satisfies both equations.

Exercise 2   Solve the system of equations

$$y = 47x - 1930$$
$$y + 19x = 710$$

by graphing. Find the intercepts of each graph to help you choose a suitable window, and use the **intersect** feature to locate the solution.

x	y
0	
	0

x	y
0	
	0

Exercise 3a.   Graph the system

$$y = -3x + 6$$
$$6x + 2y = 15$$

by hand, using either the intercept method or the slope-intercept method. (See Sections 1.1 and 1.4 to review these methods.)

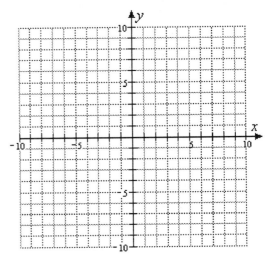

b. Identify the system as dependent, inconsistent, or consistent and independent.

Exercise 4   The manager for Books for Cooks plans to spend $300 stocking a new diet cookbook. The paperback version costs her $5, and the hardback costs $10. She finds that she will sell three times as many paperbacks as hardbacks. How many of each should she buy?

a. Let x represent the number of hardbacks and y the number of paperbacks she should buy. Write an equation about the cost of the books.

b. Write a second equation about the number of each type of book.

c. Graph both equations and solve the system. (Find the intercepts of the graphs to help you choose a window.) Answer the question in the problem.

x	y
0	
	0

x	y
0	
	0

Homework 2.1

5.

6.

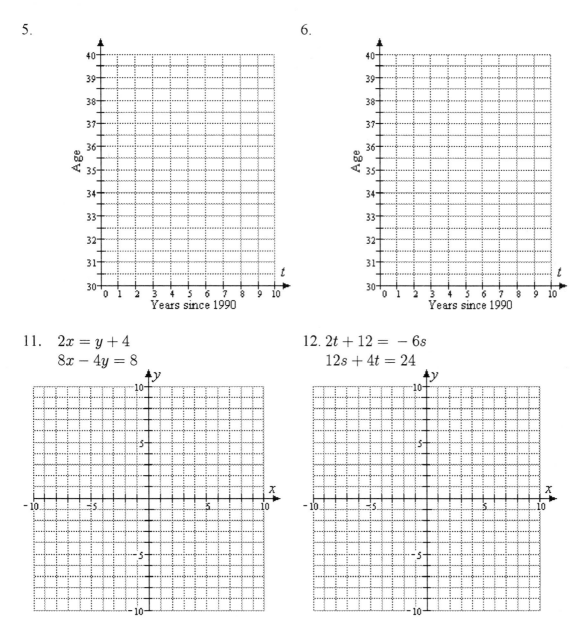

11. $2x = y + 4$
     $8x - 4y = 8$

12. $2t + 12 = -6s$
    $12s + 4t = 24$

39

13. $w - 3z = 6$
     $2w + z = 8$

14. $2u + v = 5$
     $u - 2v = 3$

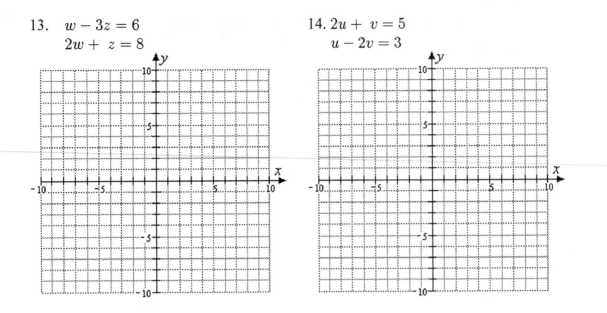

15. $2L - 5W = 6$
     $\dfrac{15W}{2} + 9 = 3L$

16. $-3A = 4B + 12$
     $\dfrac{1}{2}A + 2 = \dfrac{-2}{3}B$

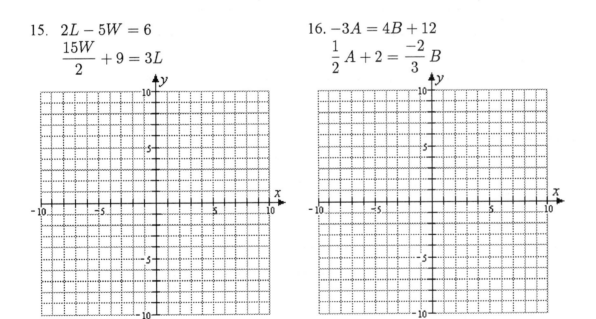

17.

x	0	30	60	90	120	150
Dash						
Friendly						

18.

x	0	6	12	18	24	30
Olympus						
Valhalla						

21.

x	5	10	15	20	25	30
C						
R						

22.

x	20	40	60	80	100	120
C						
R						

23.

	Number of tickets	Cost per ticket	Revenue
Adults	x		
Students	y		
Total			

24.

	Number of tickets	Cost per ticket	Revenue
First-class	x		
Tourist	y		
Total			

2.2 Solution of Systems by Algebraic Methods

Exercise 1   Solve the system by substitution.

$$2x - 3y = 6$$
$$x + 3y = 3$$

Exercise 2   Solve the system by linear combinations.

$$3x - 4y = -11$$
$$2x + 6y = -3$$

Exercise 3   It took Leon 7 hours to fly the same distance that Marlene drove in 21 hours. Leon flies 120 miles per hour faster than Marlene drives. At what speed did each travel?

a. Choose variables for the unknown quantities, and fill in the table.

	Rate	Time	Distance
Leon			
Marlene			

b. Write one equation about the Leon's and Marlene's speeds.

c. Write a second equation about distances.

d. Solve the system and answer the question in the problem.

Exercise 4   Identify the system as dependent, inconsistent, or consistent and independent.

$$x + 3y = 6$$
$$2x - 12 = -6y$$

Homework 2.2

23.

	Principal	Interest Rate	Interest
Bonds			
Certificate			
Total		------	

24.

	Principal	Interest Rate	Interest
First stock			
Second stock			
Total		-----	----

25.

	Pounds	% Silver	Amount of Silver
First Alloy			
Second Alloy			
Mixture			

26.

	Liters	% Acid	Amount of Acid
First Solution			
Second Solution			
Mixture			

29.

	Rate	Time	Distance
Detroit to Denver			
Denver to Detroit			

30.

	Rate	Time	Distance
Against the wind			
With the wind			

31.

	Cups	Calories per cup	Calories
Oat Flakes			
Wheat Flakes			
Mixture		----	

32.

	Sports Coupes	Wagons	Total
Hours of Riveting			
Hours of Welding			

2.3 Systems of Linear Equations in Three Variables

Exercise 1   Use back-substitution to solve the system

$$2x + 2y + z = 10$$
$$y - 4z = 9$$
$$3z = -6$$

Exercise 2   Use Gaussian reduction to solve the system

$$x - 2y + z = -1 \qquad (1)$$
$$\frac{2}{3}x + \frac{1}{3}y - z = 1 \qquad (2)$$
$$3x + 3y - 2z = 10 \qquad (3)$$

Follow the steps suggested below:
1. Clear the fractions from Equation (2).

2. Eliminate z from Equations (1) and (2).

3. Eliminate z from Equations (1) and (3).

4. Eliminate x from your new 2×2 system.

5. Form a triangular system and solve by back-substitution.

Exercise 3   Decide whether the system is inconsistent, dependent, or consistent and independent.

$$x + 3y - z = 4$$
$$-2x - 6y + 2z = 1$$
$$x + 2y - z = 3$$

Exercise 4   A manufacturer of office supplies makes three types of file cabinet: two-drawer, four-drawer, and horizontal. The manufacturing process is divided into three phases: assembly, painting, and finishing. A two-drawer cabinet requires 3 hours to assemble, 1 hour to paint, and 1 hour to finish. The four-drawer model takes 5 hours to assemble, 90 minutes to paint, and 2 hours to finish. The horizontal cabinet takes 4 hours to assemble, 1 hour to paint, and 3 hours to finish. The manufacturer employs enough workers for 500 hours of assembly time, 150 hours of painting, and 230 hours of finishing per week. How many of each type of file cabinet should he make in order to use all the hours available?

Step 1   Represent the number of each model of file cabinet by a different variable.

Number of two-drawer cabinets: x
Number of four-drawer cabinets: y
Number of horizontal cabinets: z

Step 2   Organize the information into a table. (Convert all times to hours.)

	2-Drawer	4-Drawer	Horizontal	Total Available
Assembly				
Painting				
Finishing				

Write three equations describing the time constraints in each of the three manufacturing phases. For example, the assembly phase requires $3x$ hours for the two-drawer cabinets, $5y$ hours for the four-drawer cabinets, and $4z$ hours for the horizontal cabinets, and the sum of these times should be the time available, 500 hours.

(Assembly time)   (1)

(Painting time)   (2)

(Finishing time)   (3)

Step 3   Solve the system. Follow the steps suggested below.

1. Clear the fractions from the second equation.

2. Subtract Equation (1) from 3 times Equation (3) to obtain a new Equation (4).

3. Subtract Equation (2) from twice Equation (3) to obtain a new Equation (5).

4. Equation (4) and (5) form a 2×2 system in y and z. Subtract Equation (5) from Equation (4) to obtain a new Equation (6).

5. Form a triangular system with equations (3), (4), and (6). Use back-substitution to complete the solution.

Step 4   You should have found the following solution: The manufacturer should make 60 two-drawer cabinets, 40 four-drawer cabinets, and 30 horizontal cabinets.

Midchapter Review

15.

	Rate	Time	Distance
P waves			
S waves			

16.

	Rate	Time	Distance
Thelma			
Louise			

2.4 Solution of Linear Systems Using Matrices

Exercise 1   Subtract 4 times row 2 from row 3.

$$\begin{bmatrix} 3 & 2 & -5 & | & -3 \\ 2 & -3 & 2 & | & 6 \\ 8 & -4 & 2 & | & 12 \end{bmatrix}$$

Exercise 2   Use the suggested row operations to form an equivalent matrix in upper triangular form.

$$\begin{bmatrix} 1 & -2 & 4 & | & 3 \\ 5 & -7 & 8 & | & 6 \\ -2 & 6 & -7 & | & 6 \end{bmatrix} \rightarrow \begin{bmatrix} 1 & -2 & 4 & | & 3 \\ 0 & ? & ? & | & ? \\ 0 & 0 & ? & | & ? \end{bmatrix}$$

1. Add -5(row 1) to (row 2)

2. Add 2(row 1) to (row 3)

3. Multiply (row 2) by $\frac{1}{3}$

4. Add -2(row 2) to (row 3)

Exercise 3 Use matrix reduction to solve the system

$$x + 4y = 3$$
$$3x + 8y = 1$$

Follow the suggested steps:
1. Write the augmented matrix for the system.

2. Add -3 (row 1) to (row 2).

3. Solve the resulting system by back-substitution.

Exercise 4 Use matrix reduction to solve the system

$$x + 3z = -11$$
$$2x + y + z = 1$$
$$-3x - 2y = 3$$

Follow the suggested steps.
1. Write the augmented matrix for the system.

2. Add −2(row 1) to (row 2)

3. Add 3(row 1) to (row 3)

4. Add −2(row 2) to (row 3)

5. Solve the resulting system by back-substitution.

2.5 Linear Inequalities in Two Variables

Exercise 1a. Find one y-value that satisfies the inequality

$$y - 3x < 6$$

for each of the x-values in the table.

x	1	0	-2
y			

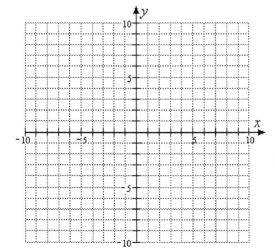

b. Graph the line

$$y - 3x = 6.$$

Then plot your solutions from part (a) on the same grid.

Exercise 2 Graph the solutions of the inequality

$$y > \frac{-3}{2}x$$

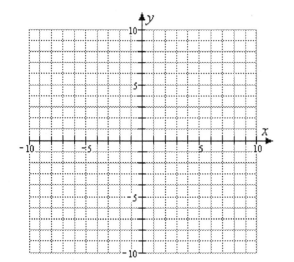

1. Graph the line $y = \frac{-3}{2}x$. (Use the slope-intercept method.)

    $$b = \qquad\qquad m = \frac{\Delta y}{\Delta x} =$$

2. Choose a test point. (Do not choose $(0, 0)$!)

3. Decide which side of the line to shade.
4. Should the boundary line be dashed or solid?

Exercise 3a. Graph the system of inequalities

$$5x + 4y < 40$$
$$-3x + 4y < 12$$
$$x < 6, \quad y > 2$$

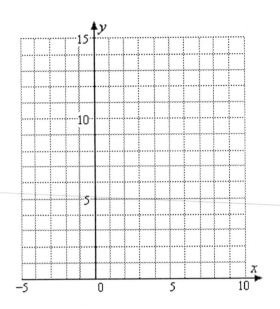

b. Find the coordinates of the vertices of the solution set.

1. Graph $\quad 5x + 4y = 40$ (Use the intercept method.)

Choose a test point. Decide which side of the line to shade.

2. Graph $\quad -3x + 4y = 12$ (Use the intercept method.)

Choose a test point. Decide which side of the line to shade.

3. Graph $\quad x = 6$.

Choose a test point. Decide which side of the line to shade.

4. Graph $\quad y = 2$.

Choose a test point. Decide which side of the line to shade.

Homework 2.5

1. $y > 2x + 4$

2. $y < 9 - 3x$

3. $3x - 2y \leq 12$

4. $2x + 5y \geq 10$

5. $x + 4y \geq -6$

6. $3x - y \leq -2$

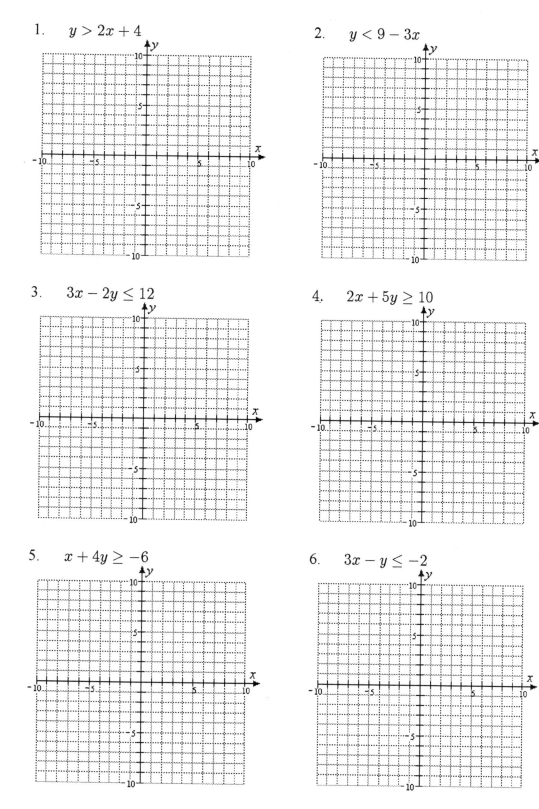

7. $x > -3y + 1$

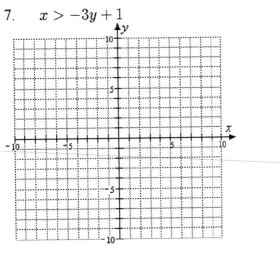

8. $x > 2y - 5$

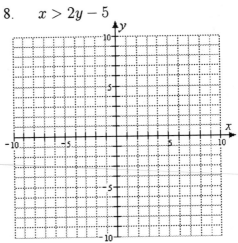

9. $x \geq -3$

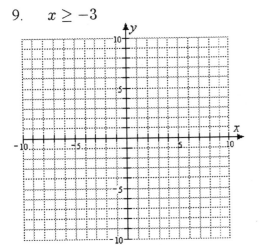

10. $y < 4$

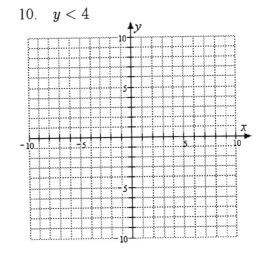

11. $y < \dfrac{1}{2}x$

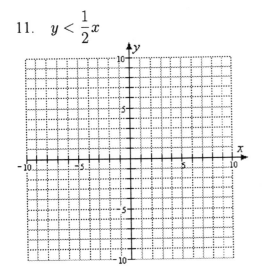

12. $y > \dfrac{4}{3}x$

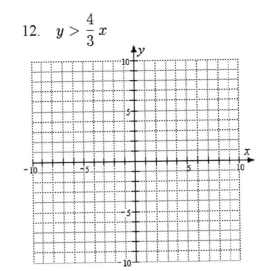

13. $0 \geq x - y$

14. $0 \geq x + 3y$

15. $-1 < y \leq 4$

16. $-2 \leq y < 0$

17. $y > 2$
    $x \geq -2$

18. $y \leq -1$
    $x > 2$

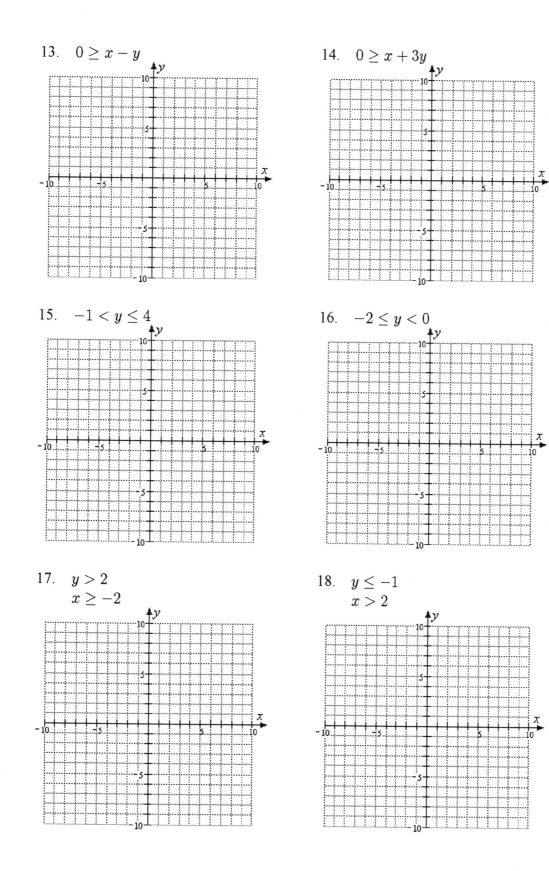

19. $y < x$
    $y \geq -3$

20. $y \geq -x$
    $y < 2$

21. $x + y \leq 6$
    $x + y \geq 4$

22. $x - y < 3$
    $x - y > -2$

23. $2x - y \leq 4$
    $x + 2y > 6$

24. $2y - x < 2$
    $x + y \leq 4$

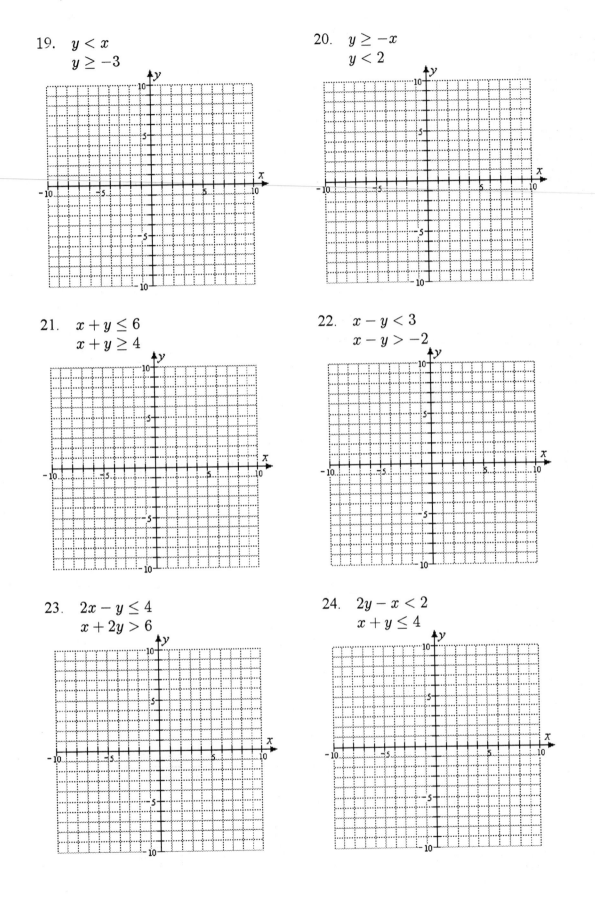

25. $3y - 2x < 2$
    $y > x - 1$

26. $2x + y < 4$
    $y > 1 - x$

27. $2x + 3y - 6 < 0$
    $x \geq 0, \ y \geq 0$

28. $3x + 2y < 6$
    $x \geq 0, \ y \geq 0$

29. $5y - 3x \leq 15$
    $x + y \leq 11$
    $x \geq 0, \ y \geq 0$

30. $y - 2x \geq -4$
    $x + y \leq 5$
    $x \geq 0, \ y \geq 0$

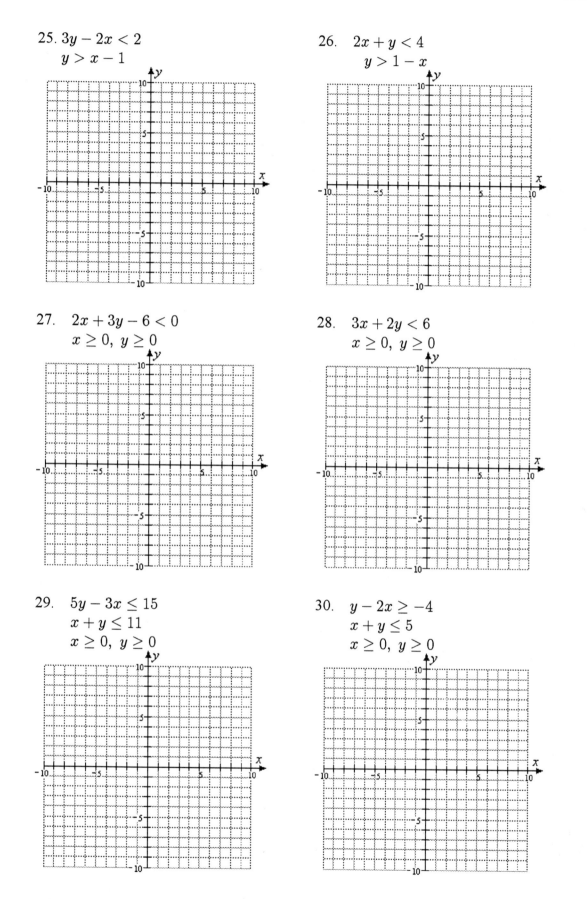

31. $2y \leq x$
    $2x \leq y + 12$
    $x \geq 0, \; y \geq 0$

32. $y \geq 3x$
    $2y + x \leq 14$
    $x \geq 0, \; y \geq 0$

33. $x + y \geq 3$
    $2y \leq x + 8$
    $2y + 3x \leq 24$
    $x \geq 0, \; y \geq 0$

34. $2y + 3x \geq 6$
    $2y + x \leq 10$
    $y \geq 3x - 9$
    $x \geq 0, \; y \geq 0$

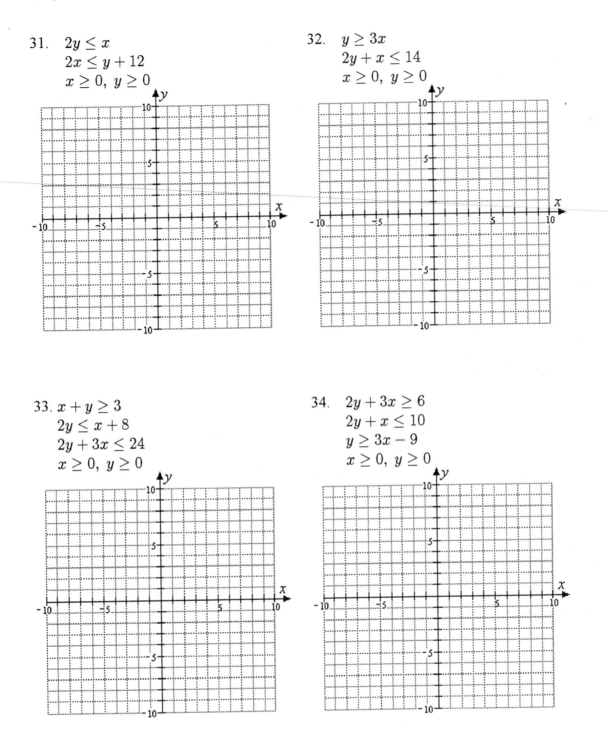

35. $3y - x \geq 3$
$y - 4x \geq -10$
$y - 2 \leq x$
$x \geq 0, \quad y \geq 0$

36. $2y + x \leq 12$
$4y \leq 2x + 8$
$x \leq 4y + 4$
$x \geq 0, \quad y \geq 0$

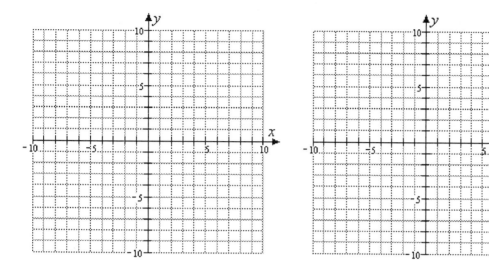

43.

44.

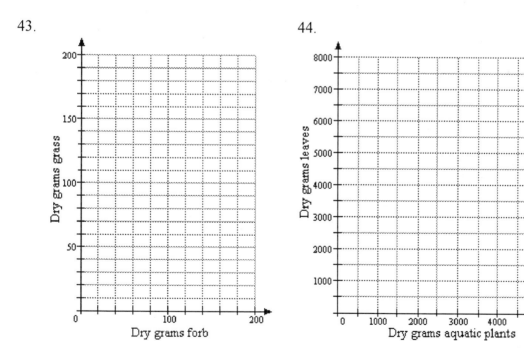

65

Chapter 2 Review

29. $3x - 4y < 12$

30. $x > 3y - 6$

31. $y < -\dfrac{1}{2}$

32. $-4 \leq x < 2$

33. $y > 3, \quad x \leq 2$

34. $y \geq x, \quad x > 2$

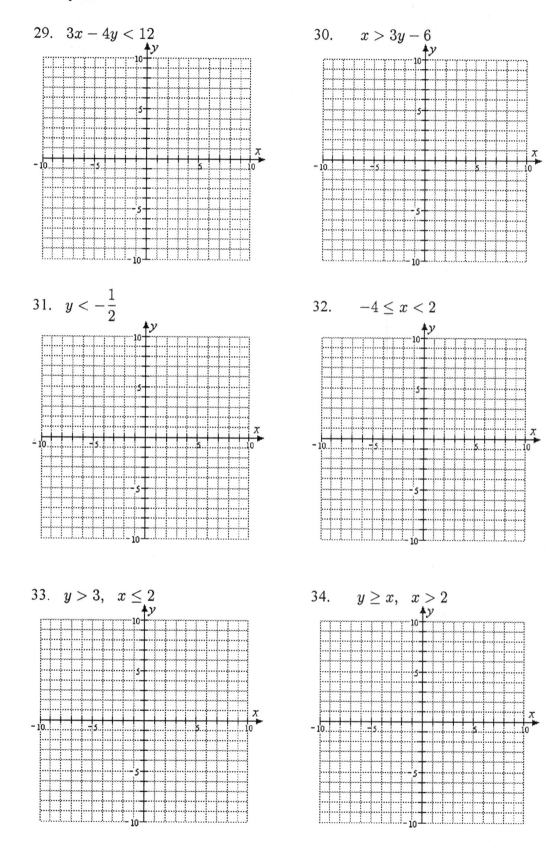

35. $3x - y < 6, \quad x + 2y > 6$

36. $x - 3y > 3, \quad y < x + 2$

37. $3x - 4y \leq 12$
    $x \geq 0, \quad y \leq 0$

38. $x - 2y \leq 6$
    $y \leq x$
    $x \geq 0, \quad y \geq 0$

39. $x + y \leq 5$
    $y \geq x$
    $y \geq 2, \quad x \geq 0$

40. $x - y \leq -3$
    $x + y \leq 6$
    $x \leq 4, \quad x \geq 0, \quad y \geq 0$

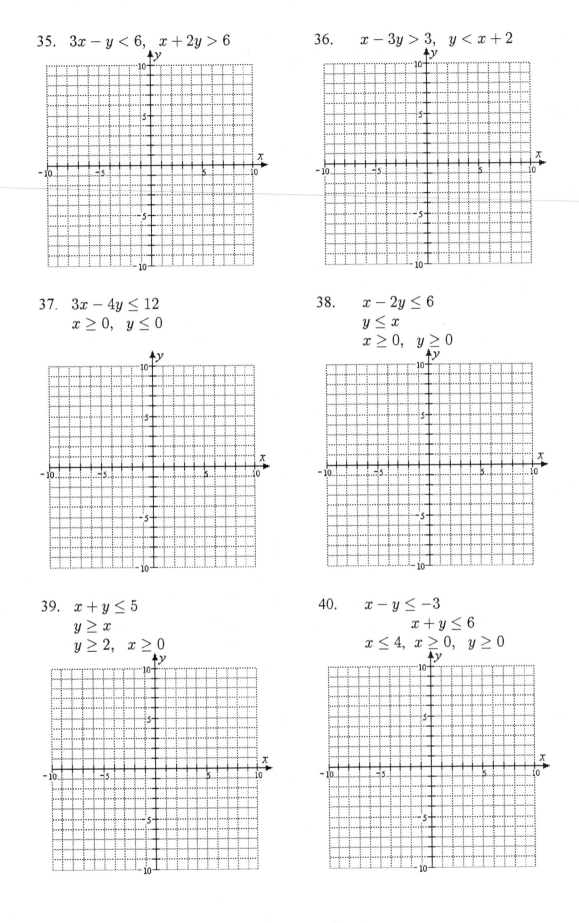

Chapter 3 Quadratic Models

Investigation 4 Falling

t	0	0.05	0.1	0.15	0.2	0.25	0.3	0.35	0.4	0.45	0.5
h											

Table 3.1

(Use the grid in Figure 3.1 on p. 71 →)

1. Complete Table 3.1 showing the distance the ball has fallen at each 0.05-second interval. Measure the distance at the bottom of the ball in each image. The first photo was taken at time $t = 0$.
2. On the gird provided, plot the points in the table. (Notice that the scale on the vertical axis increases from top to bottom instead of the usual way, from bottom to top. This is done to reflect the motion of the ball.) Connect the points with a smooth curve to sketch a graph of distance fallen versus time elapsed. Is the graph linear?
3. Use your graph to estimate the time elapsed when the ball has fallen 0.5 feet, and when it has fallen 3 feet.

4. How far did the ball fall during the first quarter second, from $t = 0$ to $t = 0.25$? How far did the ball fall from $t = 0.25$ to $t = 0.5$?

5. Add a line segment to your graph connecting the points at $t = 0$ and $t = 0.25$, and a second line segment connecting the points at $t = 0.25$ and $t = 0.5$. Compute the slope of each line segment.

6. What do the slopes in part (5) represent in terms of the problem?

7. Use your answers to part (4) to verify algebraically that the graph is not linear.

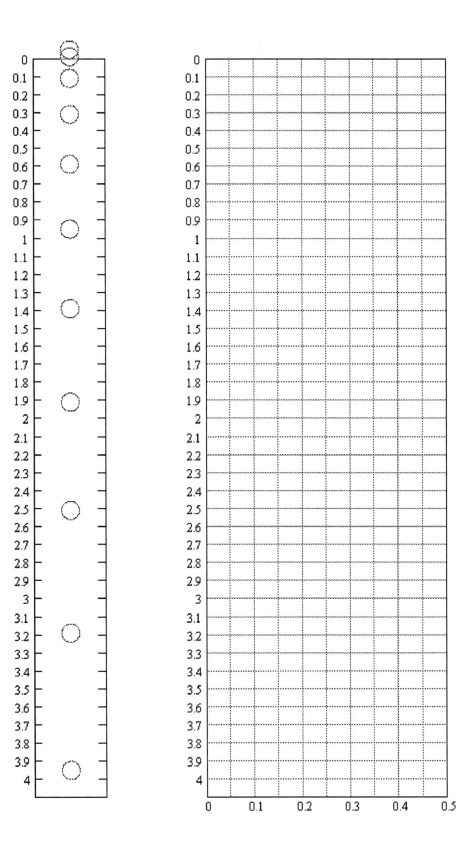

Figure 3.1

3.1 Extraction of Roots

Exercise 1   Solve by extracting roots $\dfrac{3x^2 - 8}{4} = 10$.

First, "isolate" x^2.

Take the square root of both sides.

Exercise 2   Find a formula for the radius of a circle in terms of its area.

Start with the formula for the area of a circle. $A =$

Solve for r in terms of A.

Exercise 3   Solve by extracting roots:

$$2(5x + 3)^2 = 38.$$

_a._ Give your answers as exact values.

_b._ Find approximations for the solutions to two decimal places.

Exercise 4   The average cost of dinner and a movie two years ago was $24. This year the average cost is $25.44. What was the rate of inflation over the past two years?

Homework 3.1

17.

r	1	2	3	4	5	6	7	8
V								

18.

r	1	2	3	4	5	6	7	8
V								

43.

r	0.02	0.04	0.06	0.08
B				

44.

r	0.02	0.04	0.06	0.08
A				

57.

v	0	1	2	3	4	5	6	7	8	9	10	11
J												
H												

3.2 Some Examples of Quadratic Models

Investigation 5 Perimeter and Area

Base	Height	Area
10	8	80
12	6	72
3		
14		
5		
17		
9		
2		
11		
4		
16		
15		
1		
6		
8		
13		
7		

Table 3.2

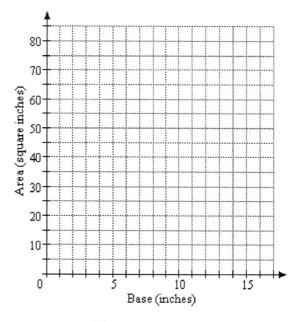

Figure 3.12

8. Base of rectangle: x

Height of rectangle:

Area of rectangle:

Investigation 6   Height of a Baseball

1. $h = -16t + 64t + 4$

t	0	1	2	3	4
h					

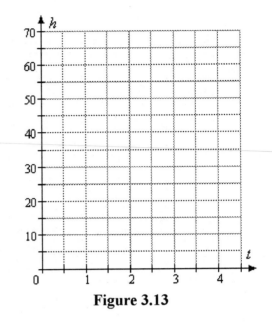

Figure 3.13

Homework 3.2

1.

Height	Base	Area	Height	Base	Area
1	34	34	10		
2	32	64	11		
3			12		
4			13		
5			14		
6			15		
7			16		
8			17		
9			18		

Height of rectangle: x

Base of rectangle:

Area of rectangle:

2.

Side	Length of Box	Width of Box	Height of Box	Volume of Box
7	1	1	3	3
8	2	2	3	12
9				
10				
11				
12				
13				
14				
15				

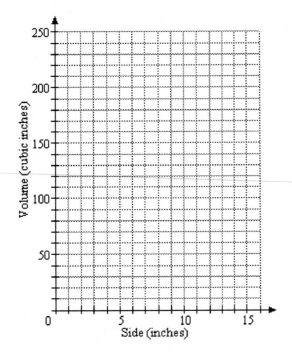

Side of cardboard sheet: x

Length of box:

Width of box:

Height of box:

Volume of box:

3.3 Solving Quadratic Equations by Factoring

Exercise 1   Graph the equation

$$y = (x - 3)(2x + 3)$$

on a calculator, and use your graph to solve the equation $y = 0$. (Use $\text{Xmin} = -9.4$, $\text{Xmax} = 9.4$.) Check your answer with the zero-factor principle.

Exercise 2   Solve by factoring $(t - 3)^2 = 3(9 - t)$.

Multiply out each side of the equation.

Obtain zero on one side of the equation.

Factor the other side.

Apply the zero-factor principle.

Exercise 3   Francine is designing the layout for a botanical garden. The plan includes a square herb garden, with a path five feet wide through the center of the garden. To include all the species of herbs, the planted area must be 300 square feet. Find the dimensions of the herb garden.

Exercise 4   Find a quadratic equation with integer coefficients whose solutions are $\frac{2}{3}$ and -5.

3.4 Graphing Parabolas: Special Cases

Exercise 1   Match each parabola with its equation. The basic parabola is shown in red.

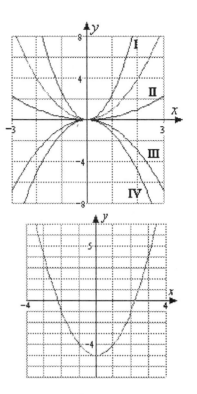

a. $y = -\frac{3}{4}x^2$
b. $y = \frac{1}{4}x^2$
c. $y = \frac{5}{2}x^2$
d. $y = -\frac{5}{4}x^2$

Exercise 2a.   Find the equation of the parabola shown.

b. Give the x- and y-intercepts of the graph.

Exercise 3a.   Find the x-intercepts and the vertex of the parabola

$$y = 6x - x^2.$$

x-intercepts: Solve $6x - x^2 = 0$

Vertex: Compute the average of the x-intercepts :

$$x_v =$$

Evaluate y at $x = x_v$:

$$y_v =$$

b. Verify your answers by graphing the equation in the window

$$\text{Xmin} = -9.4 \quad \text{Xmax} = 9.4$$
$$\text{Ymin} = -10 \quad \text{Ymax} = 10$$

81

Homework 3.4

1. $y = 2x^2$

2. $y = 4x^2$

3. $y = \dfrac{1}{2}\, x^2$

4. $y = 0.6\, x^2$

5. $y = -x^2$

6. $y = -3x^2$

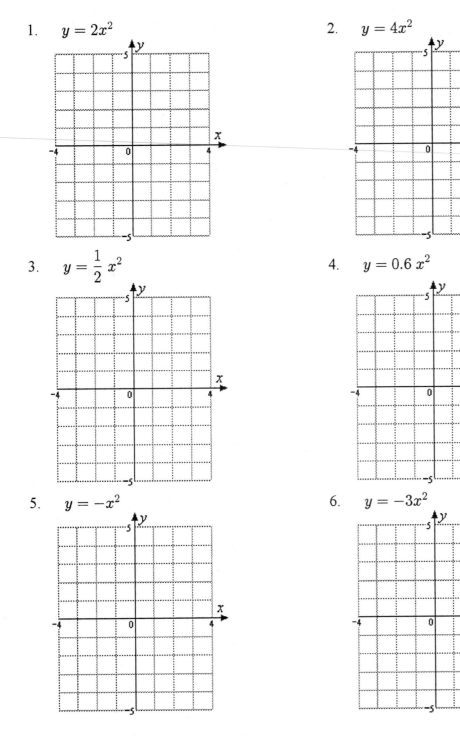

7. $y = -0.2\, x^2$

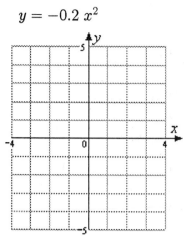

8. $y = \dfrac{-3}{4}\, x^2$

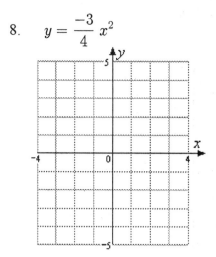

9. $y = x^2 + 2$

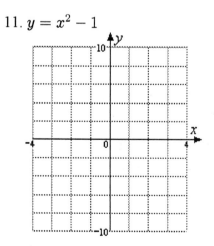

10. $y = x^2 + 5$

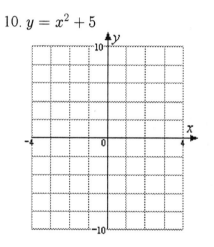

11. $y = x^2 - 1$

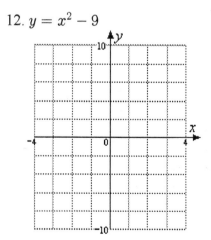

12. $y = x^2 - 9$

13. $y = x^2 - 5$

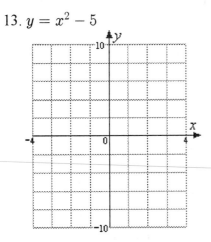

14. $y = x^2 - 3$

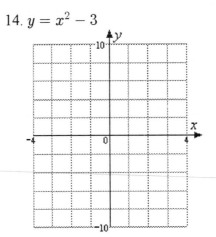

15. $y = 100 - x^2$

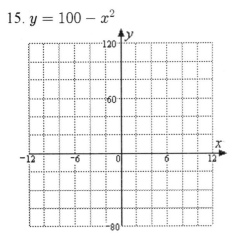

16. $y = 225 - x^2$

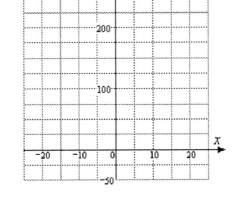
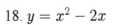

17. $y = x^2 - 4x$

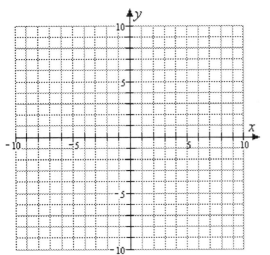

18. $y = x^2 - 2x$

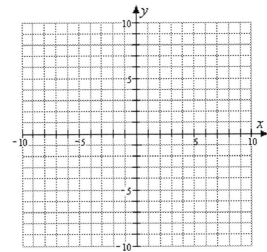

19. $y = x^2 + 2x$

20. $y = x^2 + 6x$

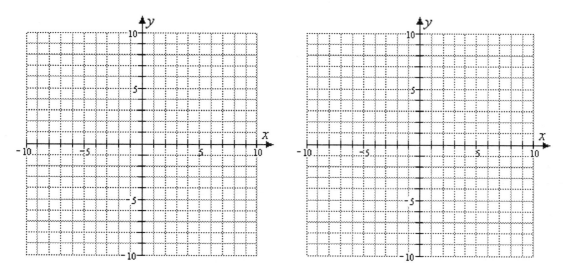

21. $y = 3x^2 + 6x$

22. $y = 2x^2 - 6x$

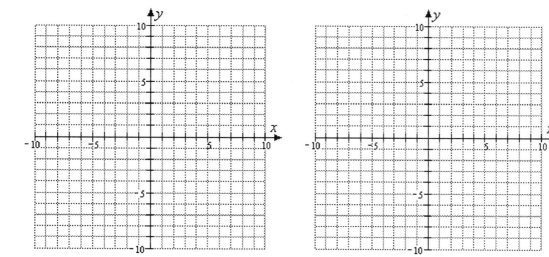

23. $y = -2x^2 + 5x$ 24. $y = -3x^2 - 8x$

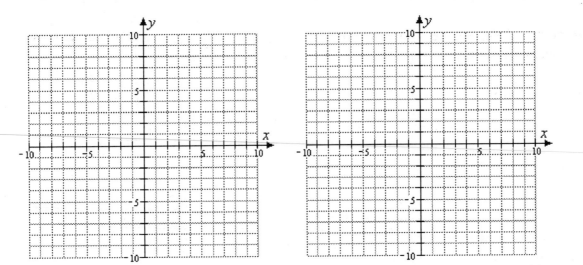

31. 32.

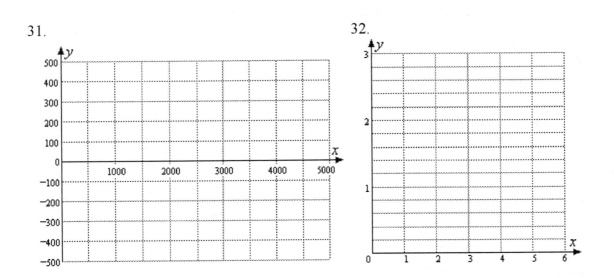

3.5   Completing the Square

Exercise 1   Complete the square by adding an appropriate constant, and write the result as the square of a binomial.

a. $x^2 - 18x + \underline{\quad} = (x \underline{\qquad})^2$

$p = \frac{1}{2}(-18) = \underline{\qquad}, \quad p^2 = \underline{\qquad}$

b. $x^2 + 9x + \underline{\quad} = (x \underline{\qquad})^2$

$p = \frac{1}{2}(9) = \underline{\qquad}, \quad p^2 = \underline{\qquad}$

Exercise 2a.   Solve by completing the square $x^2 - 1 = 3x$.

Write the equation with the constant on the right:

Complete the square on the left:

$$p = \frac{1}{2}(-3) = \underline{\quad}, \quad p^2 = \underline{\quad}$$

Add p^2 to both sides:

Write the left side as a perfect square; simplify the right side.

Solve by extracting roots:

b. Find approximations to two decimal places for the solutions.

c. Graph the parabola $y = x^2 - 3x - 1$ in the window

$$\text{Xmin} = -4.7, \quad \text{Xmax} = 4.7$$
$$\text{Ymin} = -5, \quad \text{Ymax} = 5$$

Sketch the graph on the grid.

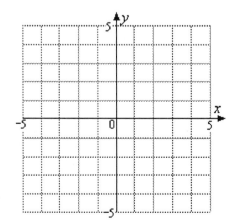

Exercise 3a. Solve by completing the square $-4x^2 - 36x - 65 = 0$.

Divide each term by -4.

Write the equation with the constant on the right:

Complete the square on the left:

$$p = \frac{1}{2}(9) = \underline{\quad}, \quad p^2 = \underline{\quad}$$

Add p^2 to both sides:

Write the left side as a perfect square;
simplify the right side.

Solve by extracting roots:

b. Graph $y = -4x^2 - 36x - 65$ in the window

$$\text{Xmin} = -9.4, \quad \text{Xmax} = 0$$
$$\text{Ymin} = -10, \quad \text{Ymax} = 20$$

3.6   Quadratic Formula

Exercise 1   Use the quadratic formula to solve $x^2 - 3x = 1$.

Write the equation in standard form.

Substitute $a = 1,\ b = -3,\ c = -1$
into the quadratic formula.

Simplify.

Exercise 2   In Investigation 6, we considered the height of a baseball, given by the equation

$$h = -16t^2 + 64t + 4.$$

Find two times when the ball is at a height of 20 feet. Give your answers to two decimal places.

Set $h = 20,$ then write the equation
in standard form.

Divide each term by -16.

Use the quadratic formula to solve.

Exercise 3   Solve $2x^2 + kx + k^2 = 1$ for x in terms of k.

Write in standard form, treating k as a constant.

Use the quadratic formula to solve.

Exercise 4   Use extraction of roots to solve $(2x + 1)^2 + 9 = 0$. Write your answers as complex numbers.

Exercise 5   Use the discriminant to determine the nature of the solutions of the equation

$$6x^2 + 13x = 240.$$

Can the equation be solved by factoring?

Write the equation in standard form, and compute the discriminant.

Chapter 3 Review

1. $y = \dfrac{1}{2}x^2$

2. $\quad y = x - 4$

3. $y = x^2 - 9x$

4. $\quad y = -2x^2 - 4x$

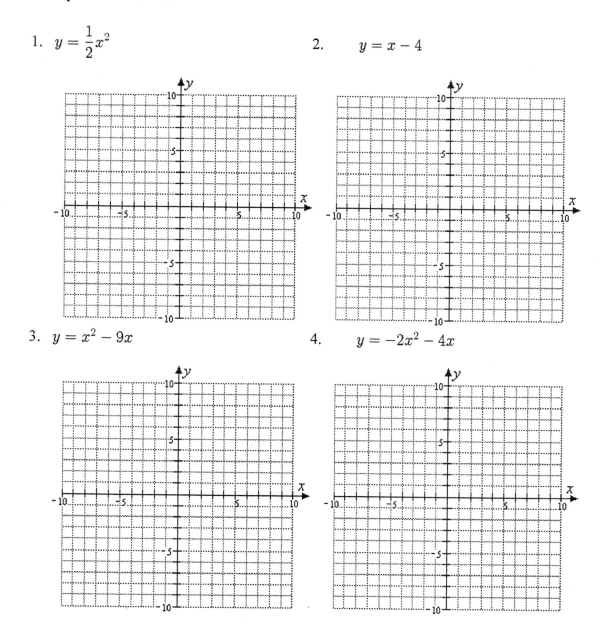

91

Chapter 4   Applications of Quadratic Models

Investigation 7   Revenue from Theater Tickets

1.

Number of Price Reductions	Price of Ticket	Number of Tickets Sold	Total Revenue
0	5.00	100	500
1	4.75	110	522.50
2			
3			
4			
5			
6			
7			
8			
9			
10			
11			

2.

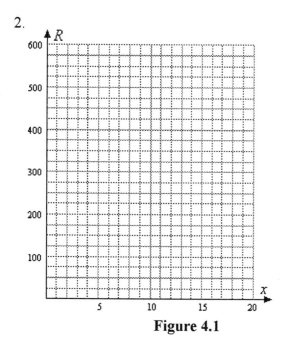

Figure 4.1

3.

Number of price reductions: $\qquad$ x

The *price of a ticket* after x price reductions:

$\quad$ *Price* =

The *number of tickets sold* at that price:

$\quad$ *Number* =

The *total revenue* from ticket sales:

$\quad$ *Revenue* =

4.1   Graphing Parabolas: The General Case

Exercise 1   Find the vertex of the graph of $y = 3x^2 - 6x + 4$. Decide whether the
vertex is a maximum point or a minimum point of the graph.

$x_v =$

$y_v =$

Exercise 2   Use the discriminant of $y = 3x^2 - 6x + 4$ to determine how many
x-intercepts the graph has.

$D = b^2 - 4ac =$

Exercise 3a. Find the intercepts and the vertex of the graph of $y = x^2 - 5x + 4$.
  b. Sketch the graph by hand.
  c. Use your calculator to verify your graph.

y-intercept: Set $x = 0$:

x-intercepts: Set $y = 0$:

vertex: $x_v =$

$y_v =$

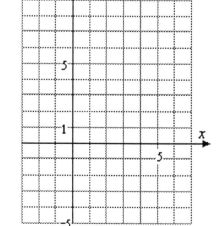

95

Exercise 4a. Find the vertex of the graph of
$y = 5 - \frac{3}{2}(x + 2)^2$.

b. Write the equation of the parabola in standard form.

c. Complete the table and sketch the graph.

x	-4	0	2
y			

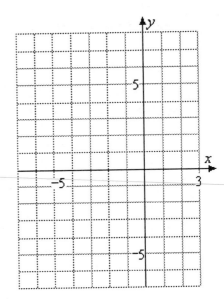

Homework 4.1

7. $y = -2x^2 + 7x + 4$

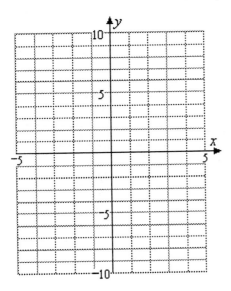

8. $y = -3x^2 + 2x + 8$

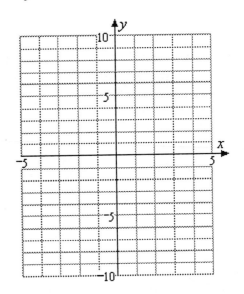

9. $y = 0.6x^2 + 0.6x - 1.2$

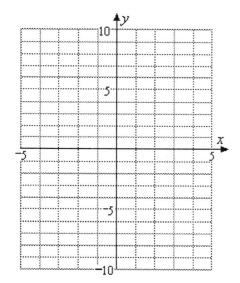

10. $y = 0.5x^2 - 0.25x - 0.75$

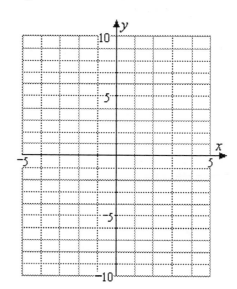

11. $y = x^2 + 4x + 7$

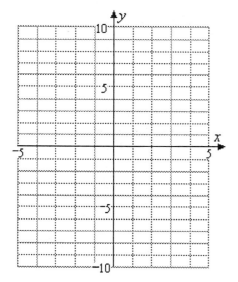

12. $y = x^2 - 6x + 10$

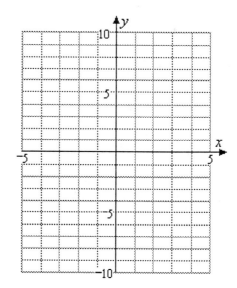

13. $y = x^2 + 2x - 1$

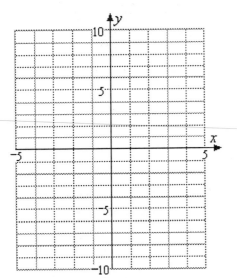

14. $y = x^2 - 6x + 2$

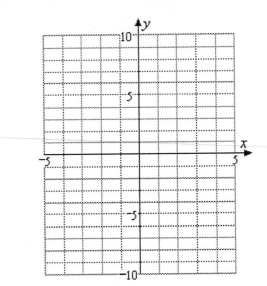

15. $y = -2x^2 + 6x - 3$

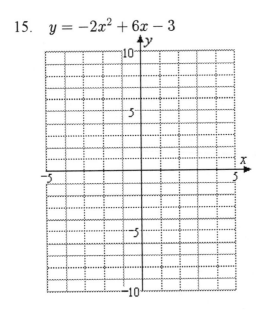

16. $y = -2x^2 - 8x - 5$

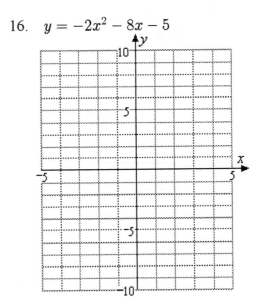

37.

t	0	0.5	1.0	1.5	2.0	2.5	3.0	3.5
x								
y								

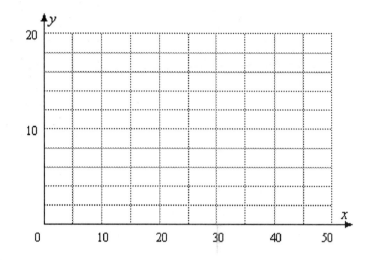

38.

Parus major

Day	1	2	3	4	5	6	7	8	9	10	11	12	13	14	15
Weight															
Growth rate															

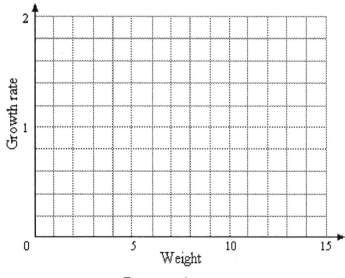

Parus major

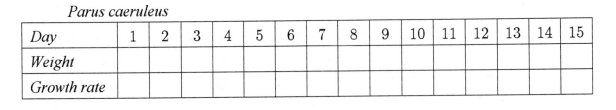

Day	1	2	3	4	5	6	7	8	9	10	11	12	13	14	15
Weight															
Growth rate															

Parus caeruleus

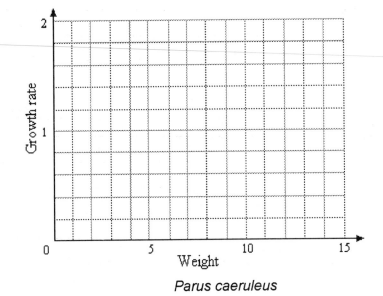

Parus caeruleus

39.

No. of price increases	Price of room	No. of rooms rented	Total revenue
0	20	60	1200
1	22	57	1254
2			
3			
4			
5			
6			
7			
8			
10			
12			
16			
20			

40.

No. of price increases	Price of tape	No. of tapes sold	Total revenue
0	6	96	576
1	6.50	92	598
2			
3			
4			
5			
6			
7			
8			
12			
16			
20			
24			

4.2   Curve Fitting

Exercise 1a. Find the equation of a parabola

$$y = ax^2 + bx + c$$

that passes through the points $(0, 80)$, $(15, 95)$, and $(25, 55)$.

b. Plot the data points and sketch the graph on the grid.

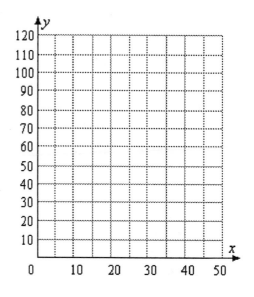

Exercise 2   Francine is designing a synchronized fountain display for a hotel in Las Vegas. For each fountain, water emerges in a parabolic arc from a nozzle three feet above the ground. Francine would like the vertex of the arc to be eight feet high, and two feet horizontally from the nozzle.

a. Choose a coordinate system and write an equation for the path of the water.

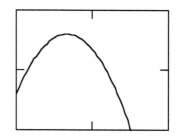

b. How far from the base of the nozzle will the stream of water hit the ground?

Exercise 3 To test the effects of radiation, a researcher irradiated male mice with various dosages and bred them to unexposed female mice. The table below shows the fraction of fertilized eggs that survived, in terms of the radiation dosage.

Radiation (rems)	100	300	500	700	900	1100	1500
Relative survival of eggs	.94	.700	.544	.424	.366	.277	.195

[Source: Strickberger, Monroe W., 1976]

a. Enter the data into your calculator and create a scatterplot. Does the graph appear to be linear? Does it appear to be quadratic?

b. Fit a quadratic regression equation to the data, and graph the equation on the scatterplot.

Homework 4.2

3.

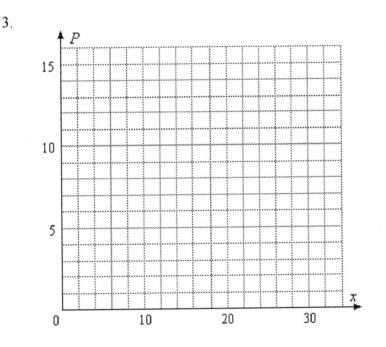

4.

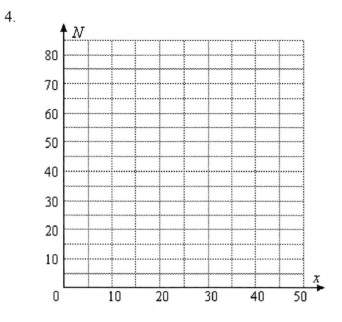

5.

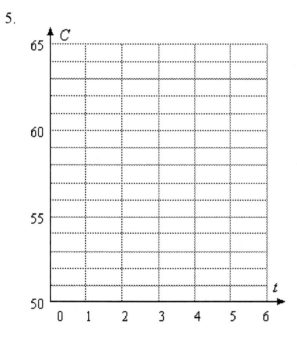

6.

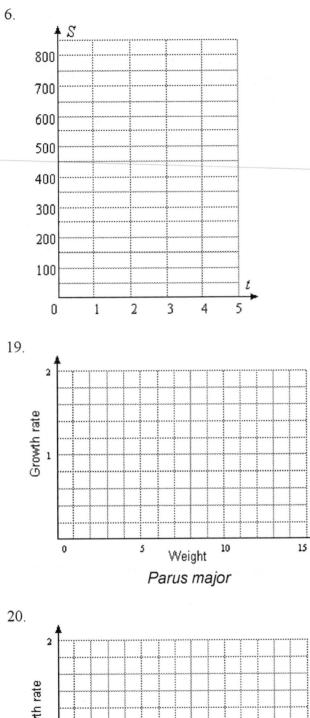

19.

20.

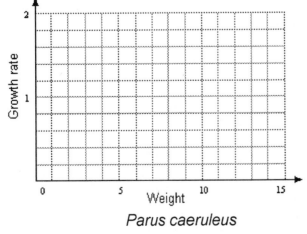

Parus caeruleus

4.3   Problem Solving

Exercise 1   The figure shows a set of three data points and a line of best fit. For this example, the regression line passes through the origin, so its equation is $y = mx$ for some positive value of m. How shall we choose m to give the best fit for the data? We want the data points to lie as close to the line as possible. One way to achieve this is to minimize the sum of the squares of the vertical distances shown in the figure.

a. The data points are $(1, 2)$, $(2, 6)$, and $(3, 7)$. Verify that the sum S we want to minimize is

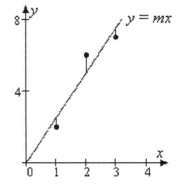

$$S = (2 - m)^2 + (6 - 2m)^2 + (7 - 3m)^2$$
$$= 14m^2 - 70m + 89.$$

b. Graph the formula for S in the window

$$\text{Xmin} = 0 \quad \text{Xmax} = 9.4$$
$$\text{Ymin} = 0 \quad \text{Ymax} = 100$$

c. Find the vertex of the graph of S.

$$m_v =$$

$$S_v =$$

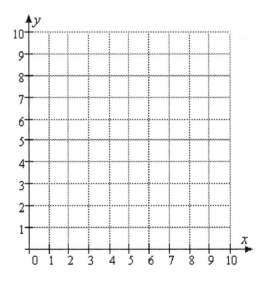

d. Use the value of m to write the equation of the regression line $y = mx$.

e. Graph the 3 data points and your regression line on the same axes.

Exercise 2a. Solve the system algebraically.

$$y = x^2 - 6x - 7$$
$$y = 13 - x^2$$

b. Graph both equations, and show the solutions on the graph.

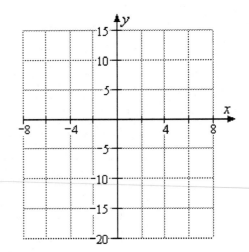

Homework 4.3

11. a.

Number of Starlings	Pecks per Minute	Pecks per Starling per Minute
1	9	
2	26	
3	48	
4	80	
5	120	
6	156	
7	175	
8	152	
9	117	
10	180	
12	132	

b.

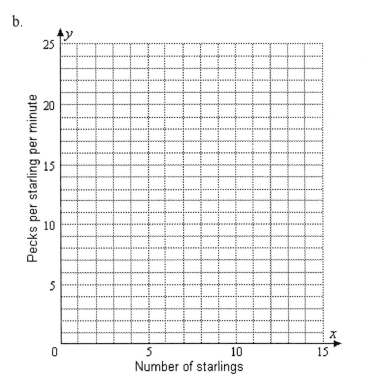

12. a.

Clutch size	2	3	4	5	6	7	8	9	19	11	12
Number of clutches	1	0	2	12	23	73	126	116	59	19	3
Number of eggs											

b.

Clutch size	1	2	3	4	5	6	7	8	9	10
Percent survival	100	90	80	70	60	50	40	30	20	10
Number of survivors										

27.

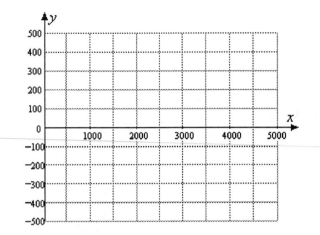

28.

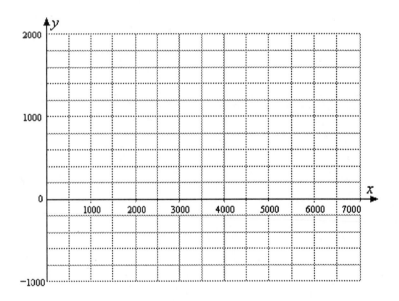

29.

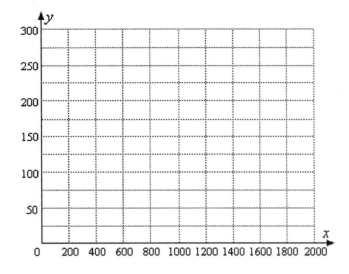

30.

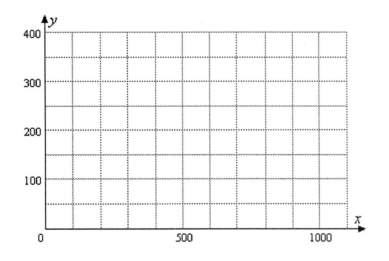

Midchapter Review

5.

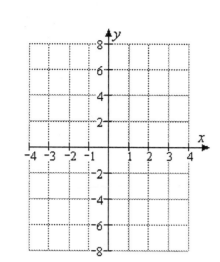

6.

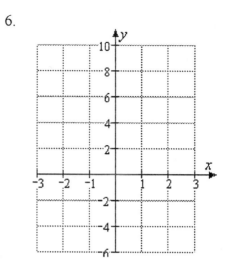

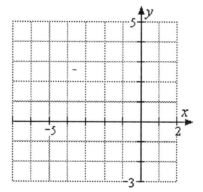

7.

x	-6	0	1
y			

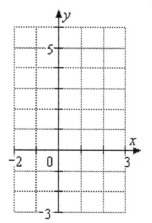

8.

x	-1	0	2
y			

113

9.

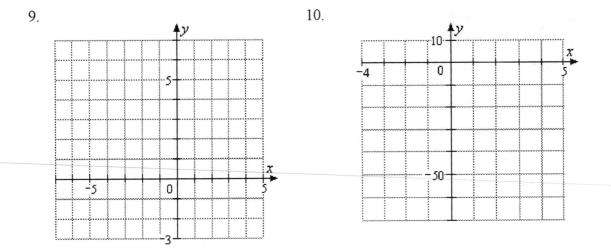

10.

17.

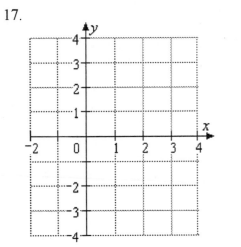

18.

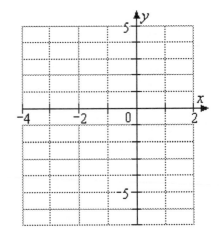

4.4 Quadratic Inequalities

Exercise 1   Use a graph to solve the inequality $x^2 - 2x - 9 \geq 6$.

Rewrite the inequality so that the right side is zero.

Sketch the graph of

$$y = x^2 - 2x - 15$$

y-intercept:

x-intercepts:

vertex:

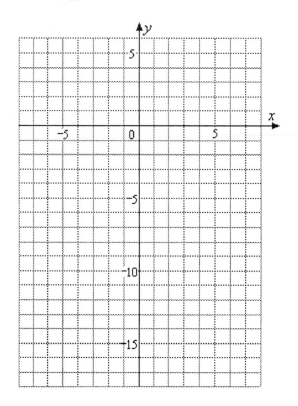

Darken the points on the graph with $y \geq 0$.
Darken the portions of the x-axis corresponding to those points.

Solution:

Exercise 2   Solve the inequality

$$36 + 6x - x^2 \leq 20.$$

Write your answer with interval notation.

a. Rewrite the inequality so that the right side is zero.

b. Graph the equation $y = 16 + 6x - x^2$.

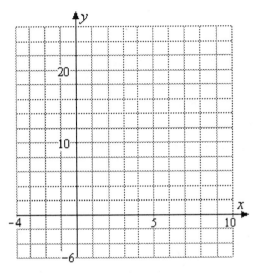

   y-intercept:

   x-intercepts: Solve $16 + 6x - x^2 = 0$.

   Vertex:

$$x_v = \frac{-b}{2a} =$$

$$y_v =$$

c. Locate the points on the graph with y-coordinate less than zero, and mark the x-coordinates of the points on the x-axis. Write the solution in interval notation.

Homework 4.4

1.

2.

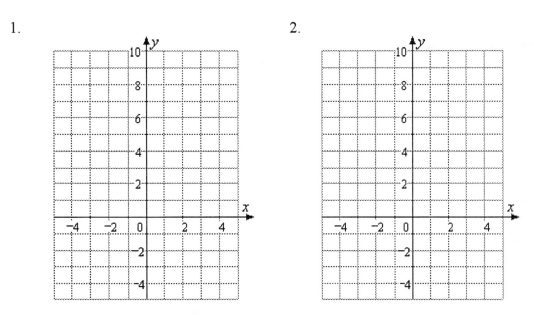

4.5 Solving Quadratic Inequalities Algebraically

Exercise 1   Solve $x^2 < 20$.

Step 1 Write the inequality in standard form.

Step 2   Find the x-intercepts of the corresponding graph. Use extraction of roots.

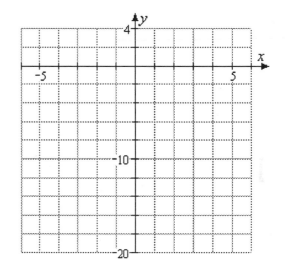

Step 3   Make a rough sketch of the graph.
Step 4   Decide which intervals on the x-axis give the correct sign for y.

Exercise 2   Solve the inequality $10 - 8x + x^2 > 4$

Step 1 Write the inequality in standard form.

Step 2 Find the x-intercepts of the corresponding graph. Use the quadratic formula.

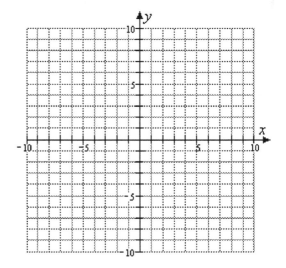

Step 3 Make a rough sketch of the graph.
Step 4 Decide which intervals on the x-axis give the correct sign for y.

   Write your answer with interval notation.

Homework 4.5

29.

Additional People	Size of Group	Price per Person	Total Income
0			
5			
10			

30.

Trees Removed	Trees per Acre	Bushels per Tree	Total Yield
0			
2			
4			

Chapter 4 Review

1. $y = x^2 - x - 12$ 2. $y = -2x^2 + x - 4$

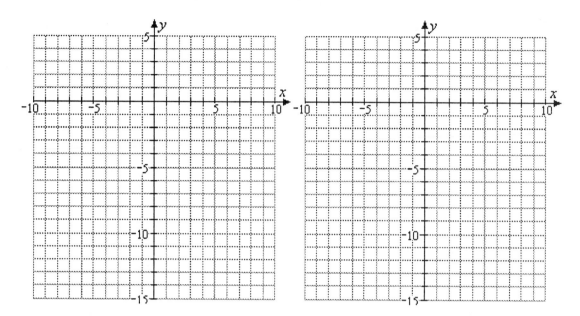

3. $y = -x^2 + 2x + 4$ 4. $y = x^2 - 3x + 4$

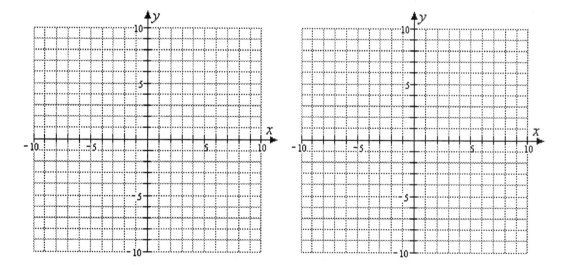

121

Chapter 5   Functions

Investigation 8   Epidemics

Day	Number Healthy	New Patients	Total Infected
0	5000	40	40
1	4960	1240	1280
2			
3			
4			
5			
6			
7			
8			
9			
10			

Table 5.1

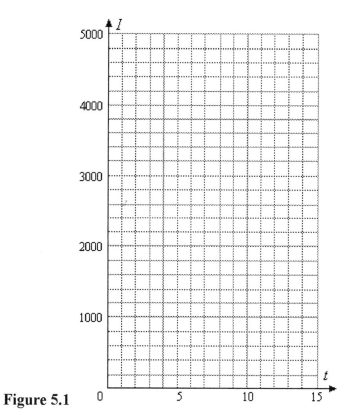

Figure 5.1

123

5.1 Definitions and Notation

Exercise 1a. As part of a project to improve the success rate of freshmen, the counseling department studied the grades earned by a group of students in English and algebra. Do you think that a student's grade in algebra is a function of his or her grade in English? Explain why or why not.

    b. Phatburger features a soda bar where you can serve your own soft drinks in any size. Do you think that the number of calories in a serving of Zap Kola is a function of the number of fluid ounces? Explain why or why not.

Exercise 2   Decide whether each table describes y as a function of x. Explain your choice.

    a.

x	3.5	2.0	2.5	3.5	2.5	4.0	2.5	3.0
y	2.5	3.0	2.5	4.0	3.5	4.0	2.0	2.5

    b.

x	-3	-2	-1	0	1	2	3
y	17	3	0	-1	0	3	17

Exercise 3   Let F be the name of the function defined by the graph in Example 5.
    a. Use function notation to state that D is a function of t.

    b. What does the statement $F(15) = 9.7$ mean in the context of the problem?

Exercise 4 When you exercise, your heart rate should increase until it reaches your target heart rate. The table shows target heart rate, $r = f(a)$, as a function of age.

a	20	25	30	35	40	45	50	55	60	65	70
r	150	146	142	139	135	131	127	124	120	116	112

a. Find $f(25)$ and $f(50)$. Does $f(50) = 2f(25)$?

b. Find a value of a for which $f(a) = 135$.

c. Find a value of a for which $f(a) = 2a$. Is $f(a) = 2a$ for all values of a?

Exercise 5 Complete the table displaying ordered pairs for the function $f(x) = 5 - x^3$.

Evaluate the function to find the corresponding $f(x)$-value for each value of x.

x	$f(x)$
-2	
0	
1	
3	

$f(-2) = 5 - (-2)^3 =$

$f(0) = 5 - 0^3 =$

$f(1) = 5 - 1^3 =$

$f(3) = 5 - 3^3 =$

Homework 5.1

35.a.

Initial clutch size laid	Experimental clutch size				
	4	5	6	7	8
5					
6					
7					
8					

b.

Initial clutch size	5	6	7	8
Optimum clutch size				

36.

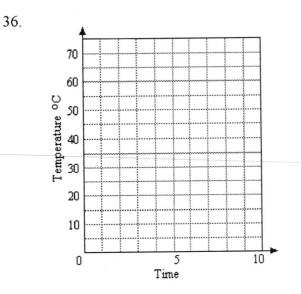

37.

Altitude (km)	Percent of Earth's Surface	Altitude (km)	Percent of Earth's Surface
−7 to −6		−1 to 0	
−6 to −5		0 to 1	
−5 to −4		1 to 2	
−4 to −3		2 to 3	
−3 to −2		3 to 4	
−2 to −1		4 to 5	

47.

t	2	5	8	10
V				

48.

x	50,000	100,000	500,000	1,000,000
S				

49.

p	5000	8000	10,000	12,000
N				

50.

t	2	5	10	20
M				

51.

d	20	50	80	100
v				

52.

h	200	1000	5000	10,000
d				

5.2 Graphs of Functions

Exercise 1 $f(x) = x^3 - 2$

    a. Complete the table of values and sketch a graph of the function.

x	-2	-1	$-\frac{1}{2}$	0	$\frac{1}{2}$	1	2
$f(x)$							

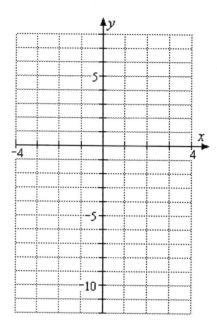

    b. Use your calculator to make a table of values and graph the function.

Exercise 2 Sketch the graph of each function.

    a. $f(x) = -2x + 5$

The graph is a line.

    Slope: $m =$

    y-intercept: $b =$

Plot the y-intercept and use the definition of slope

$$m = \frac{\Delta y}{\Delta x} =$$

to find a second point on the line.

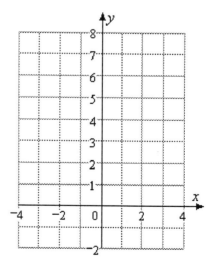

    b. $g(x) = 6 - x^2$

127

The graph is a parabola.

x-intercepts:
    Solve $0 = 6 - x^2$

Vertex:
    $x_v =$

    $y_v =$

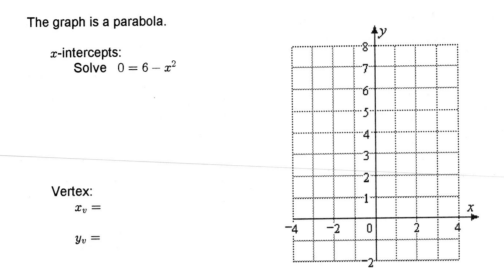

Homework 5.2

11. $g(x) = x^3 + 4;$ $x = -2, -1, \ldots, 2$

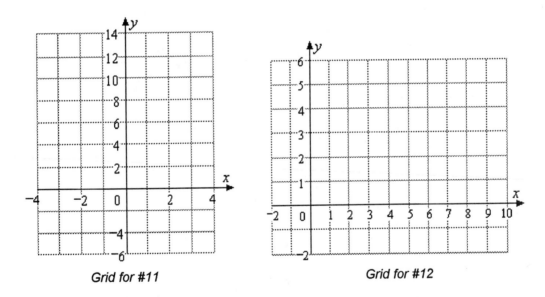

Grid for #11   Grid for #12

12. $h(x) = 2 + \sqrt{x};$ $x = 0, 1, \ldots, 9$

13. $G(x) = \sqrt{4 - x};$ $\qquad x = -5, -4, \ldots, 4$

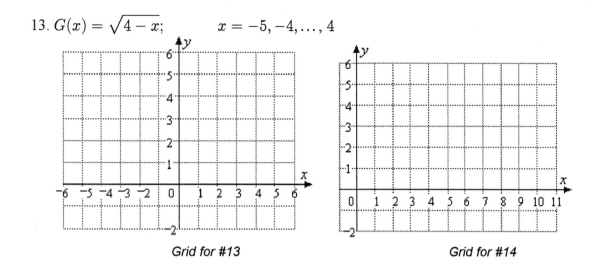

Grid for #13 $\qquad\qquad\qquad\qquad$ Grid for #14

14. $F(x) = \sqrt{x - 1};$ $\qquad x = 1, 2, \ldots, 10$

15. $v(x) = 1 + 6x - x^3;$ $\qquad x = -3, -2, \ldots, 3$

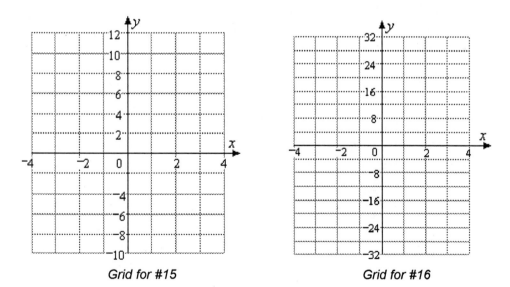

Grid for #15 $\qquad\qquad\qquad\qquad$ Grid for #16

33. $f(x) = 3x - 4$

34. $g(x) = -2x + 5$

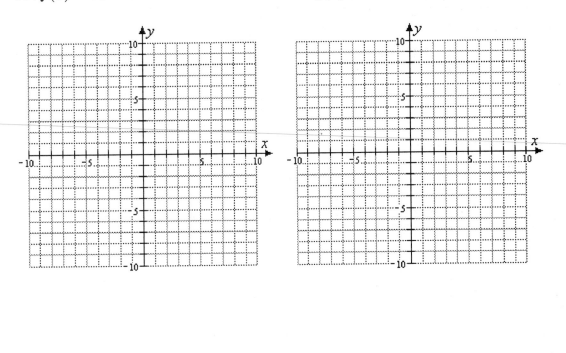

35. $G(s) = -\dfrac{5}{3}s + 50$

36. $F(s) = -\dfrac{3}{4}s + 60$

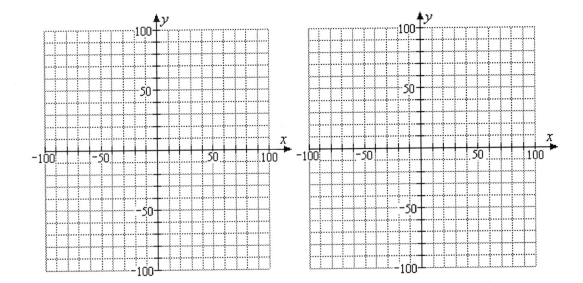

37. $h(t) = t^2 - 8$

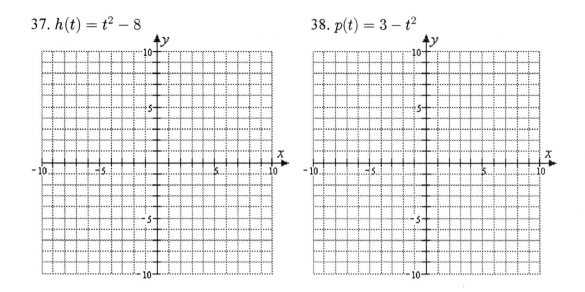

38. $p(t) = 3 - t^2$

39. $g(w) = (w + 2)^2$

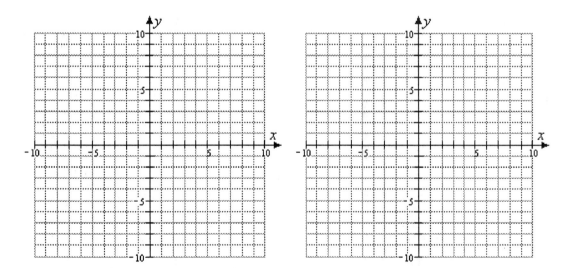

40. $f(w) = (w - 1)^2$

41. $F(z) = -2z^2$ 42. $G(z) = 0.5z^2$

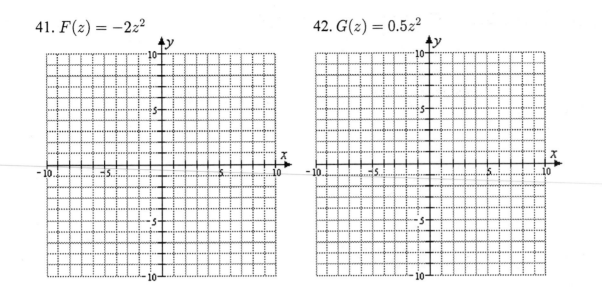

5.3   Some Basic Graphs

Exercise 1   Simplify each expression.

a. $5 - 3\sqrt[3]{64}$

b. $\dfrac{6 - \sqrt[3]{-27}}{2}$

Exercise 2   Simplify each expression.

a. $12 - 3\left| -6 \right|$

b. $-7 - 3\left| 2 - 9 \right|$

Investigation 9   Eight Basic Functions

x	$f(x) = x^2$	$g(x) = x^3$
-3		
-2		
-1		
$-\frac{1}{2}$		
0		
$\frac{1}{2}$		
1		
2		
3		

Table 5.7

133

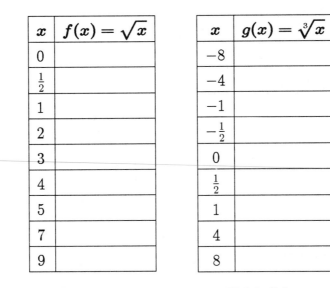

x	$f(x) = \sqrt{x}$
0	
$\frac{1}{2}$	
1	
2	
3	
4	
5	
7	
9	

Table 5.8

x	$g(x) = \sqrt[3]{x}$
-8	
-4	
-1	
$-\frac{1}{2}$	
0	
$\frac{1}{2}$	
1	
4	
8	

Table 5.9

x	$f(x) = \frac{1}{x}$	$g(x) = \frac{1}{x^2}$
-4		
-3		
-2		
-1		
$-\frac{1}{2}$		
0		
$\frac{1}{2}$		
1		
2		
3		
4		

Table 5.10

x	$f(x) = \frac{1}{x}$
-2	
-1	
-0.1	
-0.01	
-0.001	

x	$f(x) = \frac{1}{x}$
2	
1	
0.1	
0.01	
0.001	

a. **Table 5.11** b.

x	$f(x) = x$	$g(x) = |x|$
-4		
-2		
-1		
$-\frac{1}{2}$		
0		
$\frac{1}{2}$		
1		
2		
4		

Table 5.12

Homework 5.3

15. $f(x) = x^3$

16. $f(x) = |x|$

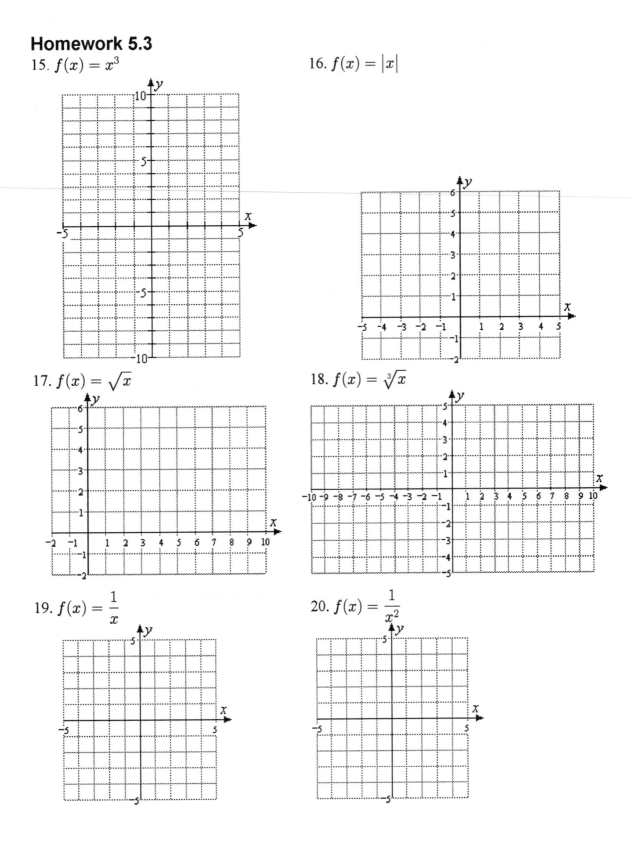

17. $f(x) = \sqrt{x}$

18. $f(x) = \sqrt[3]{x}$

19. $f(x) = \dfrac{1}{x}$

20. $f(x) = \dfrac{1}{x^2}$

Midchapter Review

9. $f(x) = 2 - \sqrt{x}$

x	0	1	2	3	4	5	6	7	8	9
$f(x)$										

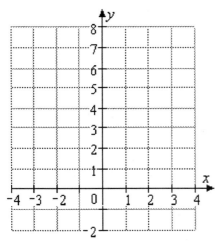

Grid for Problem 9

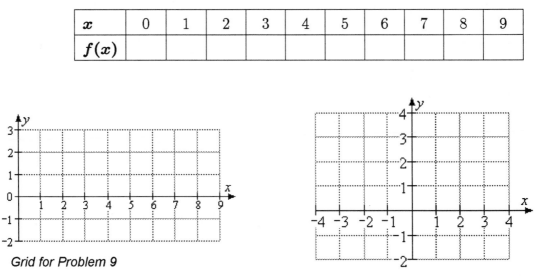

Grid for Problem 10

10. $f(x) = 3 - |x|$

x	-4	-3	-2	-1	0	1	2	3	4
$f(x)$									

11.

12.

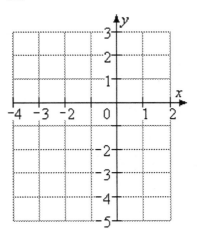

137

5.4 Domain and Range

Exercise 1a Draw the smallest viewing window possible around the graph shown.
  b. Find the domain and range of the function.

Domain:

Range:

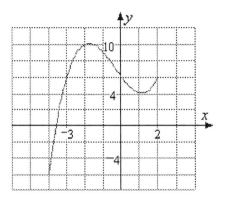

Exercise 2 Graph the function

$$g(x) = x^3 - 4$$

on the domain $[-2, 3]$ and give its range.

x	-2	-1	0	1	2	3
$f(x)$						

Range:

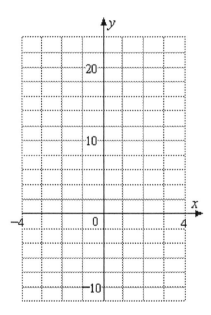

Exercise 3a. Find the domain of the function

$$h(x) = \frac{1}{(x-4)^2} \ .$$

Domain:

  b. Graph the function in the window

$$\text{Xmin} = -2, \quad \text{Xmax} = 8$$
$$\text{Ymin} = -2, \quad \text{Ymax} = 8$$

Range:

Use your graph and the function's formula to find its range.

139

5.5 Variation

Exercise 1 Which of the following graphs could represent direct variation? Explain why.

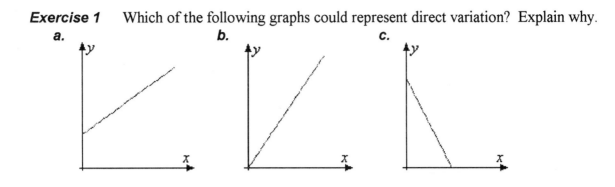

a.   b.   c.

Exercise 2 The volume of a bag of rice, in cups, is directly proportional to the weight of the bag. A two-pound bag contains 3.5 cups of rice.
  a. Express the volume, V, of a bag of rice as a function of its weight, w.

  b. How many cups of rice are in a fifteen pound bag?

Exercise 3 Delbert's office-mates want to buy a $120 gold watch for a colleague who is retiring. The cost per person is inversely proportional to the number of people who contribute.
  a. Express the cost per person, C, as a function of the number of people who contribute, p.

  b. Sketch the function on the domain $0 \le p \le 20$.

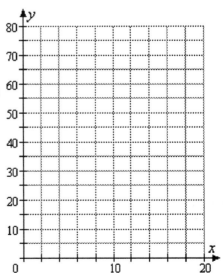

141

Homework 5.5

5.

Length	Width	Perimeter	Area
10	8		
12	8		
15	8		
20	8		

6.

Length	Width	Perimeter	Area
10	8		
12	8		
15	8		
20	8		

11.

w	100	150	200	400
m				

12.

w	100	150	200	400
m				

13.

T	1	5	10	20
L				

14.

h	2	4	5	8
L				

15.

d	2	4	12	24
F				

16.

R	1	5	10	20
I				

17.

w	10	20	40	80
P				

18.

L	0.5	1	2	4
M				

5.6 Functions as Mathematical Models

Exercise 1   Erin walks from her home to a
   convenience store, where she buys some cat
   food, and then walks back home. Sketch a
   possible graph of her distance from home as a
   function of time.

Exercise 2   Francine bought a cup of cocoa at the cafeteria. The cocoa cooled off rapidly
   at first, and then gradually approached room temperature. Which graph more accurately
   reflects the temperature of the cocoa as a function of time?   Explain why.

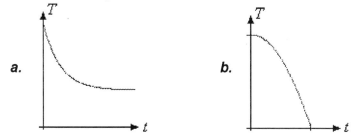

a.

b.

Exercise 3   Write each statement using absolute value notation.

   a. x is 5 units away from -3.

   b. x is at least 6 units away from 4.

143

Exercise 4a. Sketch a graph of $y = \left|2x + 7\right|$
on the grid.

x	y
-6	
-5	
-3.5	
0	

b. Use your graph to solve the inequality
$\left|2x + 7\right| < 13$.

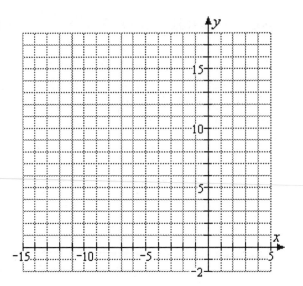

Chapter 5 Review

23. $f(t) = -2t + 4$

24. $g(s) = -\dfrac{2}{3}s - 2$

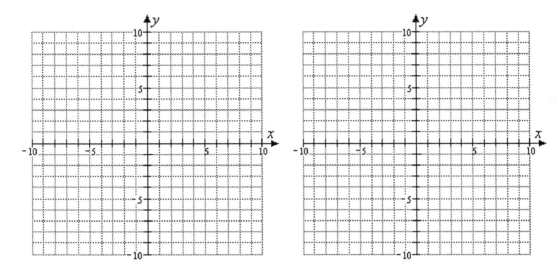

25. $p(x) = 9 - x^2$

26. $q(x) = x^2 - 16$

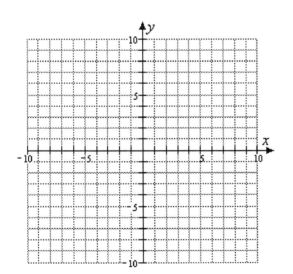

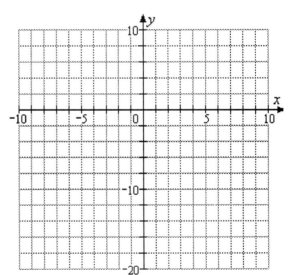

57.

x	0	4	8		16	
y				10		2

58.

x	0	4	10		14	
y				18		24

59.

x	0		4		16	25
y		1		3		

60.

x		0.5	1	1.5		4
y	4				0.5	

61.

x	-3	-2		0	1	2
y			-3			

62.

x	-3	-2		0	1	
y			8			-7

Chapter 6   Powers and Roots

Investigation 10 Inflating a Balloon

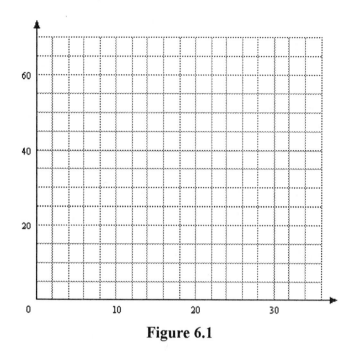

Figure 6.1

6.1 Integer Exponents

Exercise 1   Write each expression without using negative exponents.
   a. 5^{-4} ***b.*** $5x^{-4}$

Exercise 2   Solve the equation $0.2x^{-3} = 1.5$.

Rewrite without a negative exponent.

Clear the fraction.

Isolate the variable.

147

Exercise 3   Write each function as a power function in the form $y = kx^p$.

a. $f(x) = \dfrac{12}{x^2}$ **b.** $g(x) = \dfrac{1}{4x}$ **c.** $h(x) = \dfrac{2}{5x^6}$

Exercise 4a. Use the transformation $x = \dfrac{1}{r^2}$ to decide whether B is inversely proportional to r^2 for the following data.

r	0.2	0.3	0.4	0.5	0.6	0.7	0.8
B	84	45	26	12	9	8	6

First plot the data on the grid. Is the model $B = \frac{k}{r^2}$ reasonable for these data?

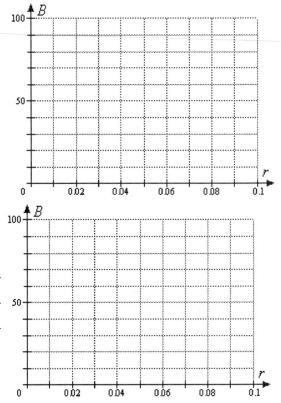

Complete the table for the transformed data points, (x, B), by computing $x = \frac{1}{r^2}$ for each given data point.

x							
B							

Plot the transformed data on the grid. Is the plot approximately linear?

b. Find a linear equation for the plot of the transformed data. Use that equation to write a power function for B in terms of r. Sketch the graph of the power function on the plot of the original data.

Draw a line of best fit through the data points. Choose two points on the line and calculate its equation.

Write a formula for B as a function of r, and sketch the graph on the grid showing the untransformed data.

Exercise 5 Write each expression without using negative exponents.

a. $\left(\dfrac{3}{b^4}\right)^{-2}$

b. $\dfrac{12}{x^{-6}}$

Exercise 6 Simplify by applying the laws of exponents.

a. $(2a^{-4})(-4a^2)$

b. $\dfrac{(r^2)^{-3}}{3r^{-4}}$

Exercise 7 Write each number in scientific notation.
a. 0.063

b. 1480

Homework 6.1

1. $f(x) = x^{-2}$

a.

x	1	2	4	8	16
$f(x)$					

c.

x	1	0.5	0.25	0.125	0.0625
x^{-2}					

2. $g(x) = x^{-3}$

a.

x	1	2	4.5	6.2	9.3
x^{-3}					

c.

x	1.5	0.6	0.1	0.03	0.002
x^{-3}					

21. a.

velocity² x $(m/s)^2$													
force F (Newtons)	12.7	15.6	19.0	22.4	23.4	24.4	30.7	27.3	32.2	30.2	36.1	37.6	35.6

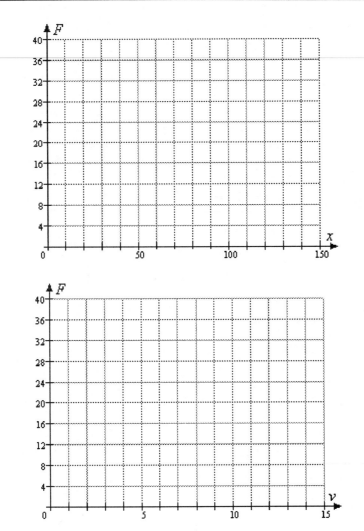

e.

22. a.

$x = \frac{1}{u}$									
$y = \frac{1}{v}$									

b.

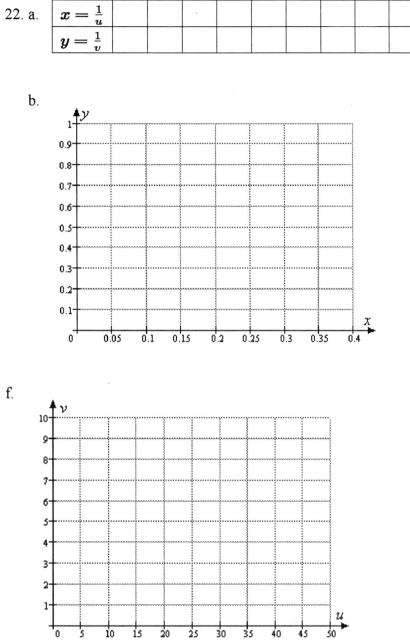

f.

151

23. a.

x, (seconds)^2						
distance, d (meters)	0.95	2.16	3.89	6.12	8.80	11.85

b.

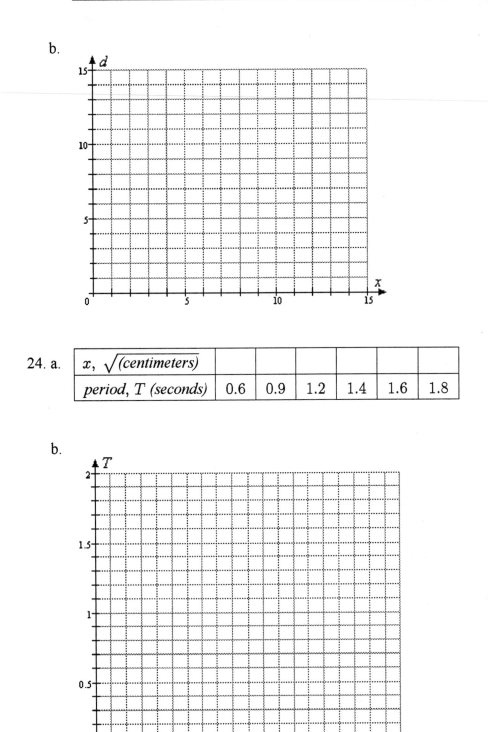

24. a.

x, $\sqrt{(centimeters)}$						
period, T (seconds)	0.6	0.9	1.2	1.4	1.6	1.8

b.

25. e.

Star	R Cygni	Betelgeuse	Arcturus	Polaris	Sirius	Rigel
λ_{max}	1.115	0.966	0.725	0.414	0.322	0.223
Temperature						

f.

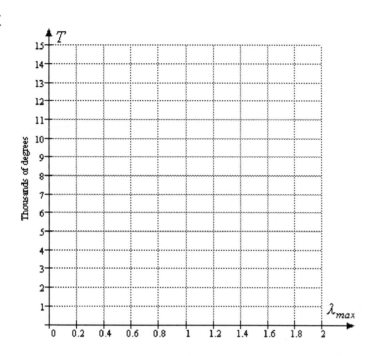

26. a.

$\frac{1}{d}$	P
	60.6
	57.2
	47.9
	38.1
	37.1
	31.9
	28.1
	26.4

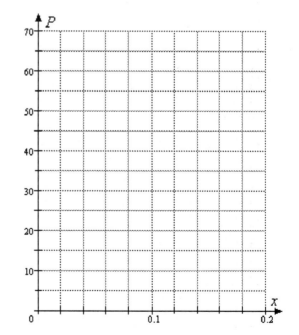

63. c.

d (nautical miles)	4	5	7	10
P (picowatts)				

e.

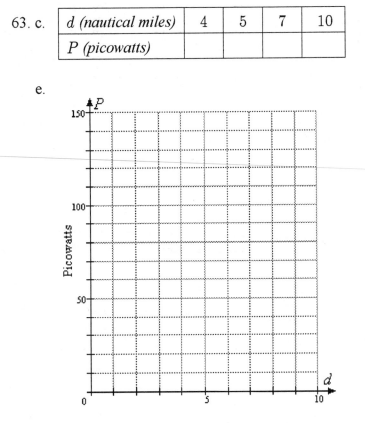

6.2 Roots and Radicals

Exercise 1   Evaluate each radical.

    a. $\sqrt[4]{16}$ **b.** $\sqrt[5]{243}$

Exercise 2   Write each power with radical notation, and then evaluate.

    a. $100{,}000^{0.2}$ **b.** $625^{1/4}$

Exercise 3a.   Convert $\dfrac{3}{\sqrt[4]{2x}}$ to exponential notation.

    b.   Convert $-5b^{0.125}$ to radical notation.

Exercise 4a.   Complete the table of values for the power function $f(x) = x^{1/4}$.

x	0	1	5	10	20	50	70	100
$f(x)$								

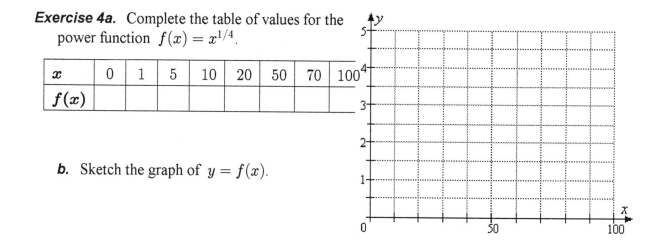

    b.   Sketch the graph of $y = f(x)$.

Exercise 5   Solve $5\sqrt[4]{x-1} = 10$.

Isolate the radical.

Raise both sides to the fourth power.

Complete the solution.

Exercise 6   Evaluate each power, if possible.

a. $-81^{1/4}$

b. $(-81)^{1/4}$

c. $-64^{1/3}$

d. $(-64)^{1/3}$

Homework 6.2

21. a.

r (meters)	0.2	0.4	0.6	0.8	1.0
v (meters per second)					

d.

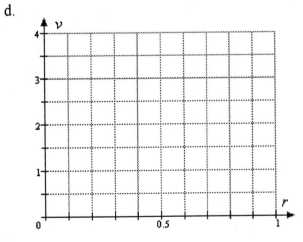

22. a.

L (feet)	200	400	600	800	1000
v_{max} (knots)					

b.

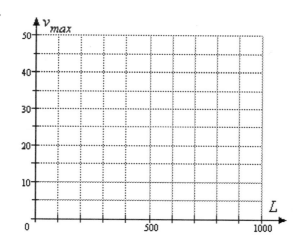

e.

L (feet)	200	400	600	800	1000
v_c (knots)					

23. c.

Element	Carbon	Potassium	Cobalt	Technetium	Radium
Mass number, A	14	40	60	99	226
Radius, r					

d.

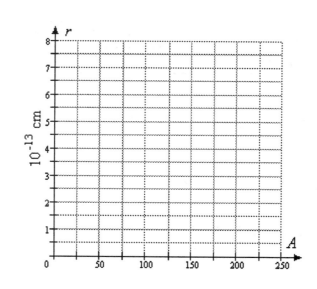

24. c.

Luminosity	10	100	1000	10,000	50,000
Mass					

d.

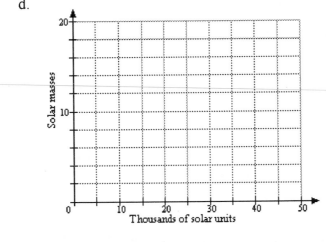

49.

a	1	2	4	8
V				

50.

v	10	20	30	40
P				

57. a.

Type of Ball	Bounce Height	e
Baseball		0.50
Basketball		0.75
Golfball		0.60
Handball		0.80
Softball		0.55
Superball		0.90
Tennisball		0.74
Volleyball		0.75

c.

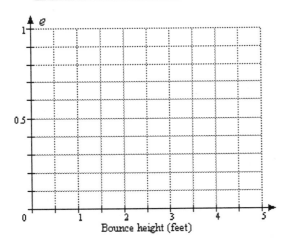

6.3 Rational Exponents

Exercise 1   Evaluate each power.

a. $32^{-3/5}$ **b.** $-81^{1.25}$

Exercise 2a. Complete the table of values for the
function $f(x) = x^{-3/4}$.

x	0.1	0.2	0.5	1	2	5	8	10
$f(x)$								

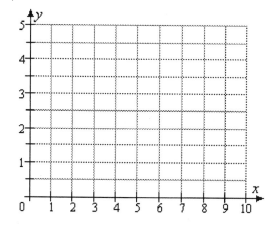

b.   Sketch the graph of the function.

Exercise 3   Convert to exponential notation.

a. $\sqrt[3]{6w^2}$

b. $\sqrt[4]{\dfrac{v^3}{s^5}}$

Exercise 4   Simplify by applying the laws of exponents.

a. $x^{1/3}(x + x^{2/3})$

b. $\dfrac{n^{9/4}}{4n^{3/4}}$

159

Exercise 5   Solve the equation $3.2z^{0.6} - 9.7 = 8.7.$ Round your answer to two decimal places.

Isolate the power.

Raise both sides to the reciprocal power.

Homework 6.3

19.

t	5	10	15	20
$I(t)$				

20.

t	6	10	14	20
$N(t)$				

25.

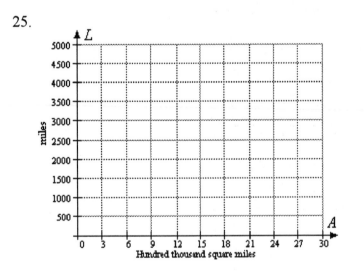

26. a.

A	10	100	1000	5000	10,000
S					

160

c.

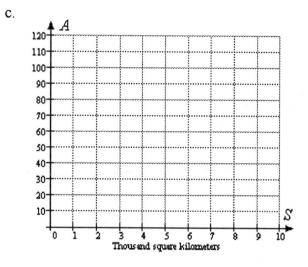

30. a.

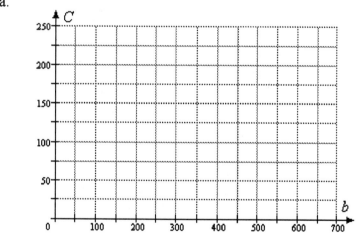

31. a.

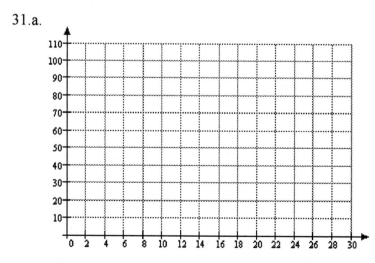

Men's Records

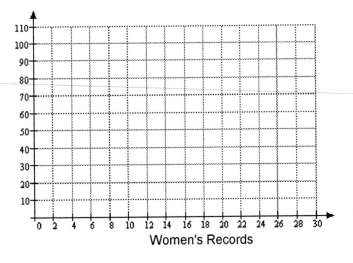

Women's Records

Midchapter Review

7. a.

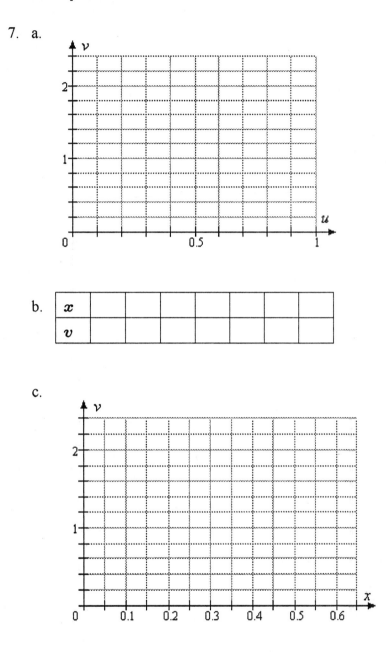

b.

x						
v						

c.

8. a.

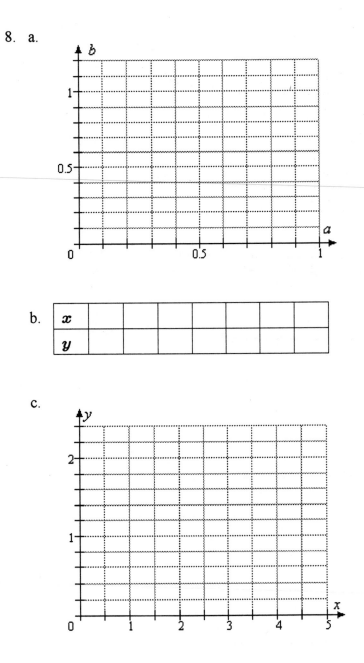

b.

x						
y						

c.

41. $f(x) = x^{0.3}$

x	0	1	5	10	20	50	70	100
$f(x)$								

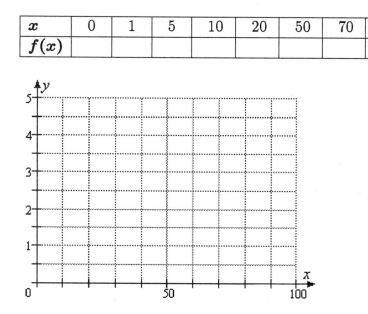

1. $f(x) = x^{-0.7}$

x	0.1	0.2	0.5	1	2	5	8	10
$f(x)$								

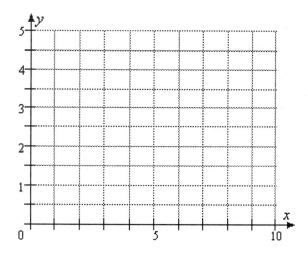

6.4   The Distance and Midpoint Formulas

Exercise 1a. Find the distance between the points $(-5, 3)$ and $(3, -9)$.

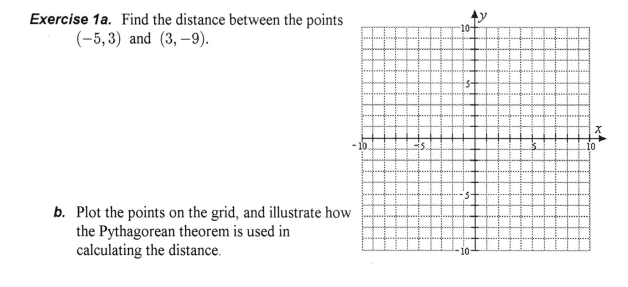

b. Plot the points on the grid, and illustrate how the Pythagorean theorem is used in calculating the distance.

Exercise 2a. Find the midpoint of the line segment joining the points $(-5, 3)$ and $(3, -9)$.

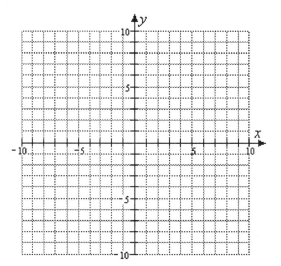

b. Plot both points and draw a rectangle with the points as opposite vertices. Illustrate that the midpoint as the center of the rectangle.

Exercise 3a. State the center and radius of the circle

$$(x+3)^2 + (y+2)^2 = 16$$

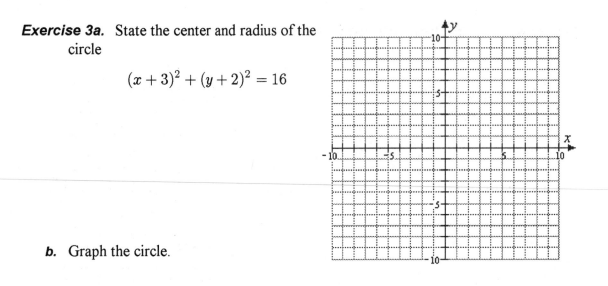

b. Graph the circle.

Exercise 4 Write the equation of the circle

$$x^2 + y^2 - 14x + 4y + 25 = 0$$

in standard form.

Homework 6.4

7.

8.

9.

10.

11.

12.

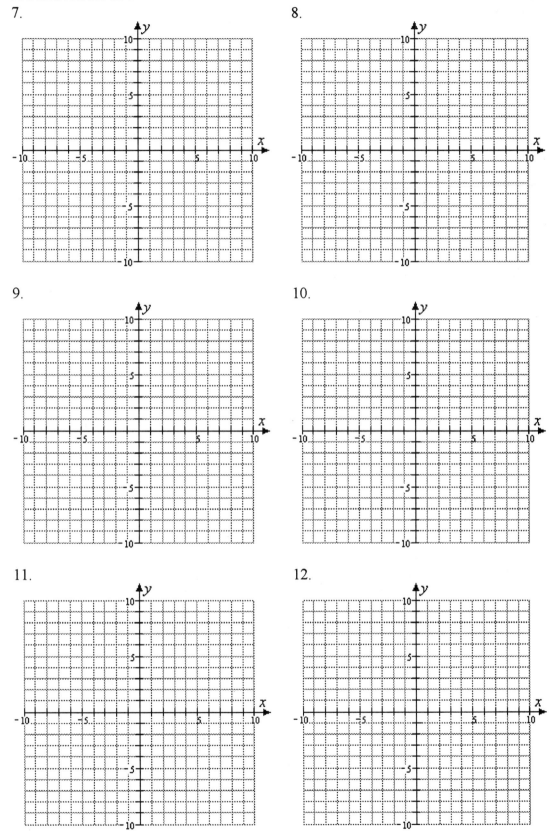

13.

14.

15.

16.

17.

18.

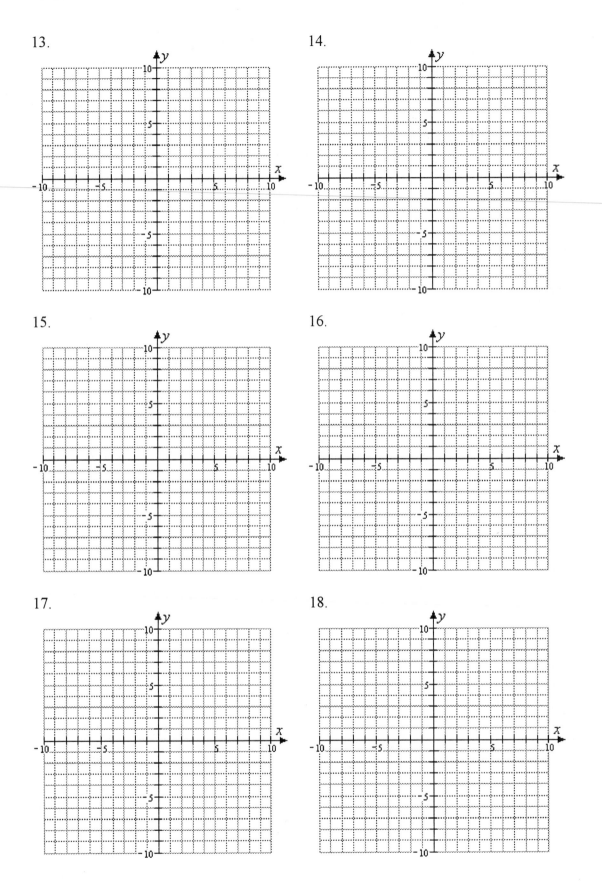

19.

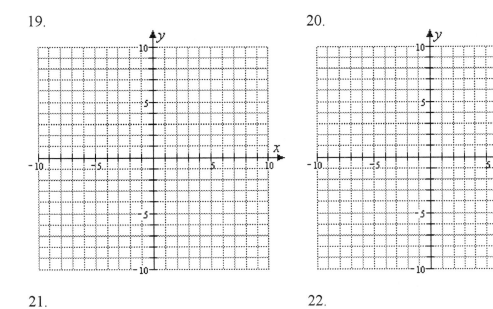

20.

21.

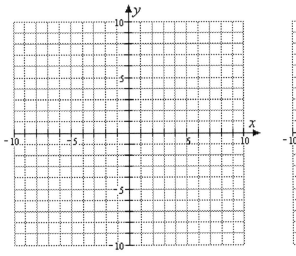

22.

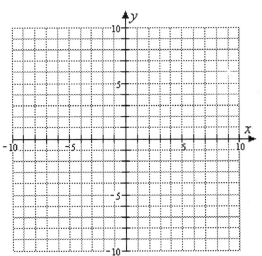

23.

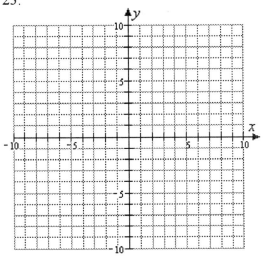

24.

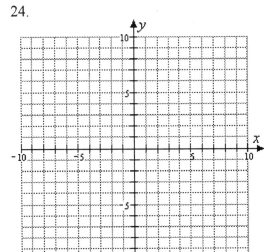

25.

26.

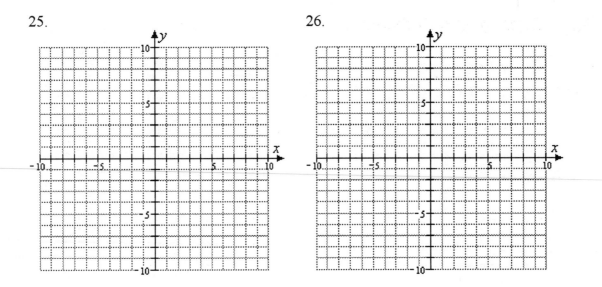

6.5 Working with Radicals

Exercise 1   Simplify $\sqrt[3]{250b^7}$.

Look for perfect cube factors of $250b^7$.

Apply Property (1).

Take cube roots.

Exercise 2   Simplify.

a. $\sqrt{\dfrac{18x^5}{25y^4}}$

b. $\sqrt[3]{2a^2}\,\sqrt[3]{6a^2}$

Exercise 3   Simplify $\sqrt[3]{40x^2} - 3\sqrt[3]{16x^2} + \sqrt[3]{54x^2}$.

Exercise 4   Expand $(\sqrt{5} - 2\sqrt{3})^2$.

173

Exercise 5   Rationalize the denominator of $\dfrac{-\sqrt{3}}{\sqrt{7}}$.

Exercise 6   Rationalize the denominator of $\dfrac{\sqrt{3}}{\sqrt{3} - \sqrt{2}}$.

6.6 Radical Equations

Exercise 1   Solve $6 + 2\sqrt[4]{12 - v} = 10$.

Isolate the radical.

Raise each side to the fourth power.

Complete the solution.

Exercise 2   Solve $2x - 5 = \sqrt{40 - 3x}$.

Square both sides.

Solve the quadratic equation.

Check for extraneous roots.

Exercise 3   Solve the formula $r - 2 = \sqrt[3]{V - Bh}$ for h.

175

Exercise 4   Solve $\sqrt{3x+1} = 6 - \sqrt{9-x}$.

Square both sides.

Isolate the radical term.

Divide both sides by 4.

Square both sides again.

Solve the quadratic equation.

Check for extraneous roots.

Chapter 6 Review

10. a.

Planet	Radius (km)	Mass (10^{20} kg)	Density (kg/m³)
Mercury	2440	3302	
Venus	6052	48,690	
Earth	6378	59,740	
Mars	3397	6419	
Jupiter	71,490	18,990,000	
Saturn	60,270	5,685,000	
Uranus	25,560	866,200	
Neptune	24,765	1,028,000	
Pluto	1150	150	

11. a.

v^2, $(m/s)^2$					
Energy (cal/kg sec), E	0.9	1.8	2.6	3.9	5

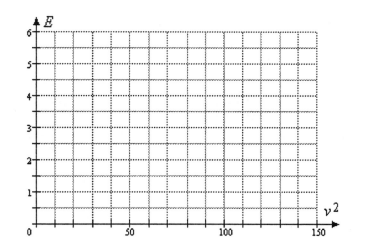

12. a.

$\dfrac{1}{P}$				
Heating time (minutes), T	14.5	13	10	9

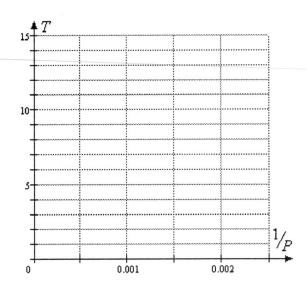

53.

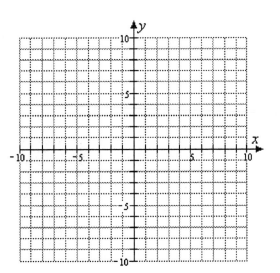

54.

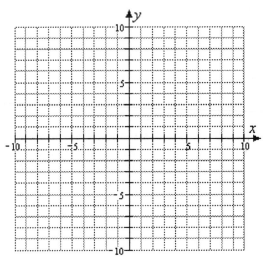

55.

56.

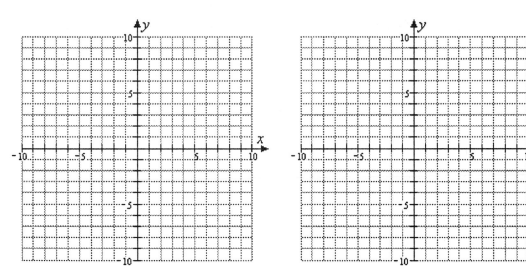

Chapter 7 Exponential and Logarithmic

7.1 Exponential Growth and Decay

Investigation 11   Population Growth

A.

t	$P(t)$
0	100
1	
2	
3	
4	
5	

$P(0) = 100$

$P(1) = 100 \cdot 3 =$

$P(2) = [100 \cdot 3] \cdot 3 =$

$P(3) =$

$P(4) =$

$P(5) =$

Table 7.1

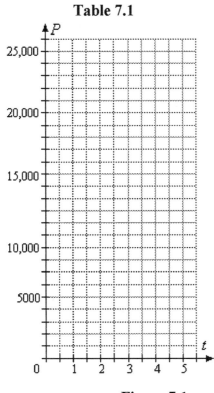

Figure 7.1

B.

t	$P(t)$
0	60
3	
6	
9	
12	
15	

$P(0) = 60$

$P(3) = 60 \cdot 2 =$

$P(6) = [60 \cdot 2] \cdot 2 =$

$P(9) =$

$P(12) =$

$P(15) =$

Table 7.2

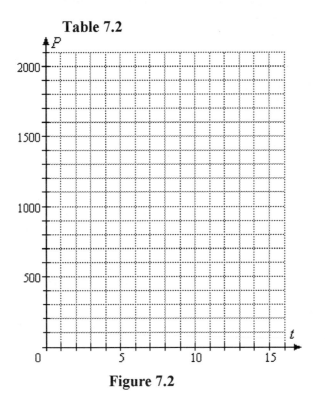

Figure 7.2

Exercise 2   In 1998, the average annual cost of a public college was \$10,069, and costs were climbing by 6% per year.

a.   Complete the table and sketch a graph of $C(t)$.

t	0	5	10	15	20	25
$C(t)$						

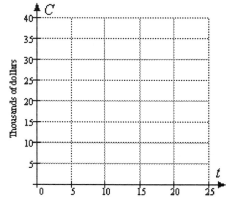

b.   Write a formula for $C(t)$, the cost of one year of college t years after 1998.

c.   If the percent growth rate remains steady, how much would a year of college cost in 2001?

d.   How much would a year of college cost in 2020?

Investigation 12   Exponential Decay

A.

t	$P(t)$
0	5000
10	
20	
30	
40	
50	

$P(0) = 5000$

$P(10) = 5000 \cdot 0.90 =$

$P(20) = [5000 \cdot 0.90] \cdot 0.90 =$

$P(30) =$

$P(40) =$

$P(50) =$

Table 7.6

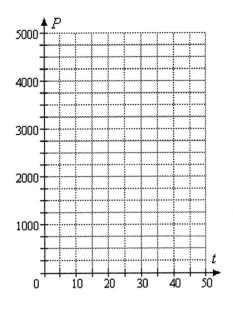

Figure 7.4

B.

x	$P(x)$
0	100
1	
2	
3	
4	
5	

Table 7.7

$P(0) = 100$

$P(1) = 100 \cdot 0.75 =$

$P(2) = [100 \cdot 0.75] \cdot 0.75 =$

$P(3) =$

$P(4) =$

$P(5) =$

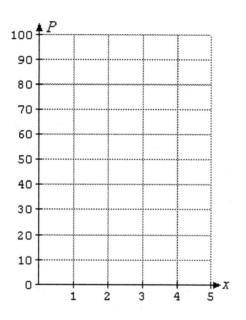

Figure 7.5

Exercise 3   The number of butterflies visiting a nature station is declining by 18% per year. In 1998, 3600 butterflies visited the nature station.

     a. What is the decay factor in the annual butterfly count?

     b. Write a formula for $B(t)$, the number of butterflies t years after 1998.

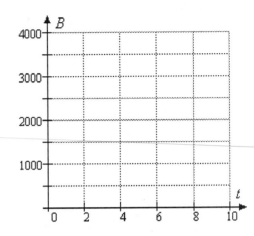

     c. Complete the table and sketch a graph of $B(t)$.

t	0	2	4	6	8	10
$B(t)$						

Homework 7.1

1.

Weeks	0	1	2	3	4
Bacteria					

2.

Months	0	1	2	3	4
Fruit Flies					

3.

Weeks	0	6	12	18	24
Bees					

4.

Years	0	3	6	9	12
Cattle					

5.

Years	0	1	2	3	4
Account Balance					

6.

Years	0	1	2	3	4
Account Balance					

7.

Years since 1963	0	1	2	3	4
Value of House					

8.

Years since 1990	0	1	2	3	4
Windsurfers					

9.

Weeks	0	2	4	6	8
Mosquitoes					

10.

Years	0	5	10	15	20
Perch					

11.

Feet	0	4	8	12	16
% of Light					

12.

Years	0	3	6	9	12
Value of Boat					

13.

Years	0	1	2	3	4
Pounds of Plutonium-238					

14.

Days	0	1	2	3	4
Grams of Iodine-131					

15.

t	0	1	2
P(t)			
Q(t)			

16.

t	0	1	2
P(t)			
Q(t)			

19. a.

n	$-\infty$	1	2	3	4	5	6
r(n)							

b.

Planet	Mercury	Venus	Earth	Mars	Jupiter	Saturn
Orbital Radius (AU)	0.39	0.72	1.00	1.52	5.20	9.54
n						

Table 7.10

20. a.

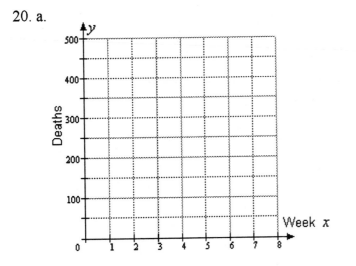

b.

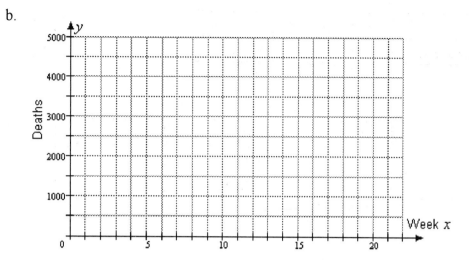

d.

Week	Total Deaths	Week	Total Deaths
0, May 9	9	12, August 1	
1, May 16		13, August 8	
2, May 23		14, August 15	
3, May 30		15, August 22	
4, June 6		16, August 29	
5, June 13		17, September 5	
6, June 20		18, September 12	
7, June 27		19, September 19	
8, July 4		20, September 26	
9, July 11		21, October 3	
10, July 18		22, October 10	
11, July 25			

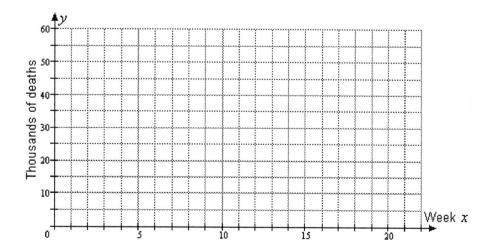

189

21.

t	P
0	8
1	12
2	18
3	
4	

22.

t	P
0	4
1	5
2	6.25
3	
4	

23.

x	Q
0	20
1	24
2	
3	
4	

24.

x	Q
0	100
1	105
2	
3	
4	

25.

w	N
0	120
1	96
2	
3	
4	

26.

w	N
0	640
1	480
2	
3	
4	

41.

t	0	2	4	6	8
L(t)					

t	0	2	4	6	8
E(t)					

42.

t	0	3	6	9	12
L(t)					

t	0	3	6	9	12
E(t)					

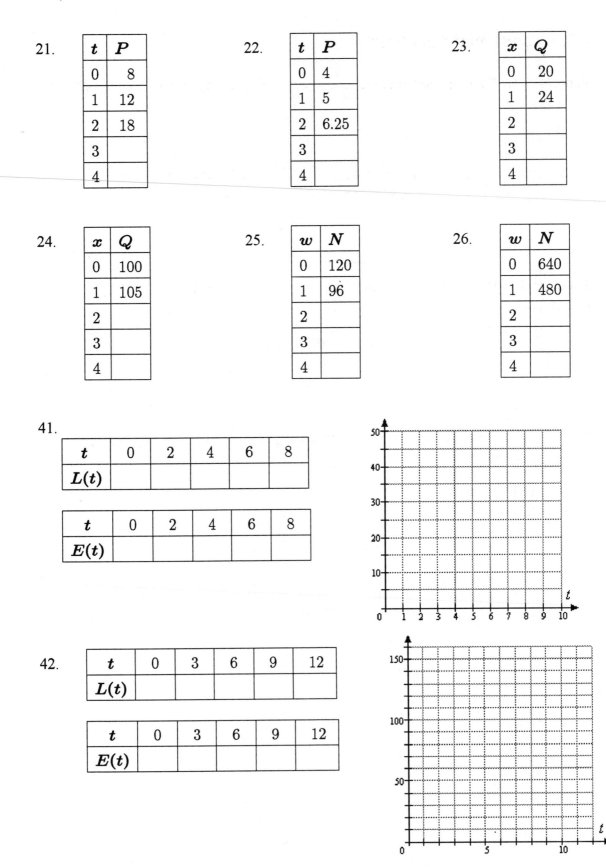

190

7.2 Exponential Functions

Exercise 1a. State the ranges of the functions f and g from Example 1 on the domain $[-2, 2]$.

b. State the ranges of the functions p and q shown in Figure 7.10 on the domain $[-2, 2]$. Round your answers to two decimal places.

Exercise 2 Which of the functions below have the same graph? Explain why.

a. $f(x) = \left(\frac{1}{2}\right)^x$ **b.** $g(x) = -2^x$ **c.** $h(x) = 2^{-x}$

Exercise 3 Which of the following functions are exponential functions, and which are power functions?

a. $F(x) = 1.5^x$ **b.** $G(x) = 3x^{1.5}$
c. $H(x) = 3^{1.5x}$ **d.** $K(x) = (3x)^{1.5}$

Exercise 4 Solve the equation $2^{x+2} = 128$.

Write each side as a power of 2.

Equate exponents.

Exercise 5   During an advertising campaign in a large city, the makers of Chip-O's corn chips estimate that the number of people who have heard of Chip-O's increases by a factor of 8 every 4 days.

   a. If 100 people are given trial bags of Chip-O's to start the campaign, write a function $N(t)$ for the number of people who have heard of Chip-O's after t days of advertising.

   b. Use your calculator to graph the function $N(t)$ on the domain $0 \le t \le 15$.

   c. How many days should they run the campaign in order for Chip-O's to be familiar to 51,200 people? Use algebraic methods to find your answer, and verify on your graph.

Exercise 6   Use the graph of $y = 5^x$ to find an approximate solution to $5^x = 285$, accurate to two decimal places.

Homework 7.2

39.b.

Order	Number	Average Length	Total Length
1	1,600,000	1	
2	339,200	2.3	
3			
4			
5			
6			
7			
8			
9			
10			

40.

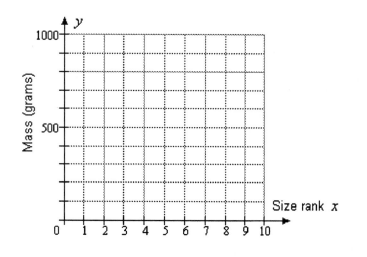

41. a.

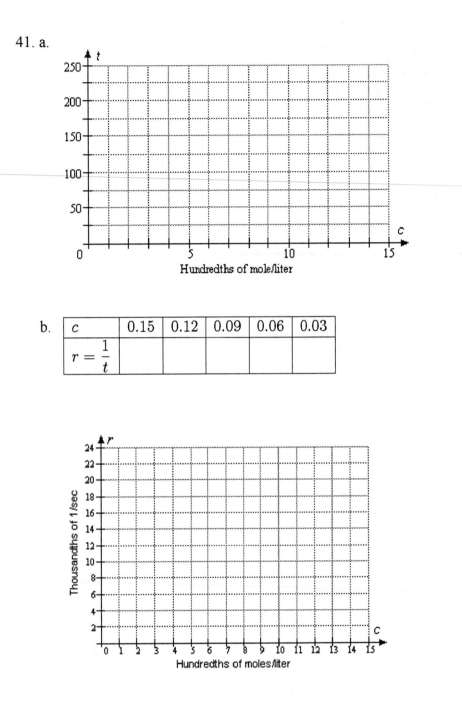

b.

c	0.15	0.12	0.09	0.06	0.03
$r = \dfrac{1}{t}$					

42. a.

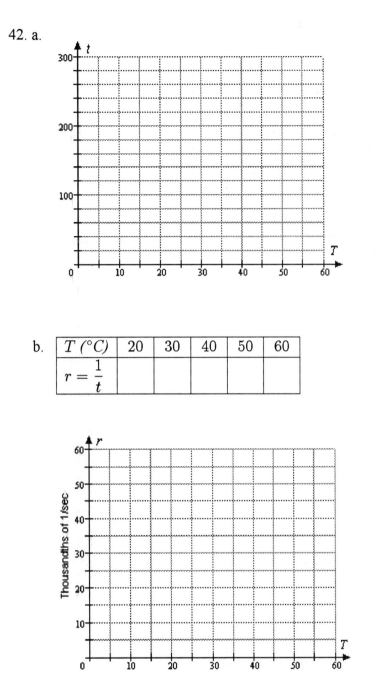

b.

$T\ (^{\circ}C)$	20	30	40	50	60
$r = \dfrac{1}{t}$					

43. a.

Stage n	S(n)	N(n)	P(n)
0			
1			
2			
3			

44. a.

Stage n	S(n)	A(n)	N(n)	R(n)	T(n)
1					
2					
3					

45.

x	$f(x) = x^2$	$g(x) = 2^x$
-2		
-1		
0		
1		
2		
3		
4		
5		
6		

46.

x	$f(x) = x^3$	$g(x) = 3^x$
-2		
-1		
0		
1		
2		
3		
4		
5		
6		

47. $f(x) = 2^{x-1}$, $g(x) = 2^x - 1$

x	$y = 2^x$	$f(x)$	$g(x)$
-2			
-1			
0			
1			
2			

48. $f(x) = 3^x + 2$, $g(x) = 3^{x+2}$

x	$y = 3^x$	$f(x)$	$g(x)$
-2			
-1			
0			
1			
2			

49. $f(x) = -3^x$, $g(x) = 3^{-x}$

x	$y = 3^x$	$f(x)$	$g(x)$
-2			
-1			
0			
1			
2			

50. $f(x) = 2^{-x}$, $g(x) = -2^x$

x	$y = 2^x$	$f(x)$	$g(x)$
-2			
-1			
0			
1			
2			

7.3 Logarithms

Exercise 1   Find each logarithm.

    a. $\log_3 81$
                                      b. $\log_{10} \frac{1}{1000}$

Exercise 2   Rewrite each equation in logarithmic form.

    a. $8^{-1/3} = \dfrac{1}{2}$
                          b. $5^x = 46$

Exercise 3a. Rewrite the equation $3^x = 90$ in logarithmic form.

    b. Use a graph to approximate the solution to the equation in part (a). Round your answer to three decimal places.

Exercise 4   Solve $20 + 10^x = 220$.

Isolate the power of 10.

Rewrite in logarithmic form.

Evaluate.

Exercise 5   The percentage of American homes with computers grew exponentially from 1994 to 1999. For $t = 0$ in 1994, the growth law was $P(t) = 25.85(10)^{0.052t}$.

 a. What percent of American homes had computers in 1994?

 b. If the percentage of homes with computers continues to grow at the same rate, when will 90% of American homes have a computer?

 c. Do you think that the function $P(t)$ will continue to model the percentage of American homes with computers? Why or why not?

Exercise 6   Solve for x: $\log_4 x = 3$

Midchapter Review

3.

Years since 1992	0	1	2	3	4
Value of house					

4.

Years since 1990	0	1	2	3	4
Computers sales					

5.

Years since 1995	0	1	2	3	4
Agriculture majors					

6.

Years since 1998	0	1	2	3	4
Value of SUV					

7. $f(x) = 1.8^x$

8. $g(x) = (0.65)^x$

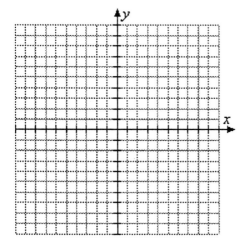

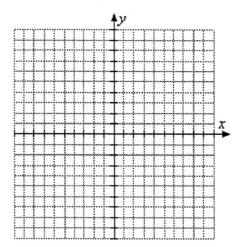

201

7.4 Logarithmic Functions

Exercise 1   Simplify each expression.

a. $\log_{10} 10^6$

b. $10^{\log_{10} 1000}$

c. $\log_4 4^6$

d. $4^{\log_4 64}$

Exercise 2   Evaluate $t = \dfrac{1}{k} \log_{10} \dfrac{C_H}{C_L}$ for $k = 0.05$, $C_H = 2$, and $C_L = 0.5$.

Exercise 3   The pH of the water in a tide pool is 8.3. What is the hydrogen ion concentration of the water?

Exercise 4   The noise of city traffic registers at about 70 decibels.
  a. What is the intensity of traffic noise, in watts per square meter?

  b. How many times more intense is traffic noise than conversation?

Homework 7.4

1. $f(x) = 2^x$

x	$f(x)$
-2	
-1	
0	
1	
2	

x	$g(x)$

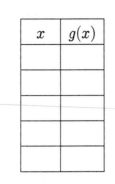

2. $f(x) = 3^x$

x	$f(x)$
-2	
-1	
0	
1	
2	

x	$g(x)$

3. $f(x) = \left(\frac{1}{3}\right)^x$

x	$f(x)$
-2	
-1	
0	
1	
2	

x	$g(x)$

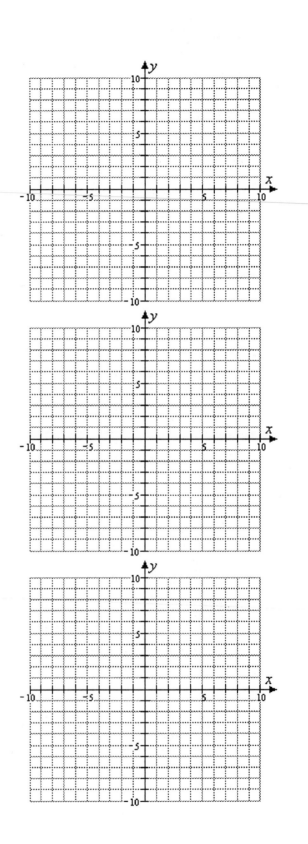

4. $f(x) = \left(\frac{1}{2}\right)^x$

x	$f(x)$
-2	
-1	
0	
1	
2	

x	$g(x)$

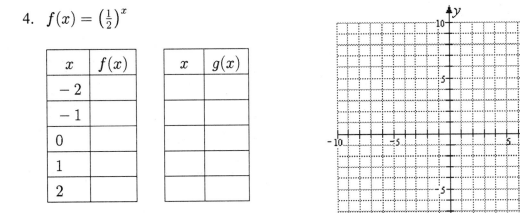

35. a.

Distance from bed (feet)	0.2	0.4	0.6	0.8	1.0	1.2
Velocity, Hoback River (ft/sec)						
Velocity, Pole Creek (ft/sec)						

c.

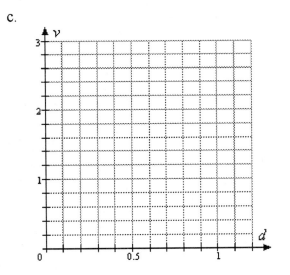

205

36. a.

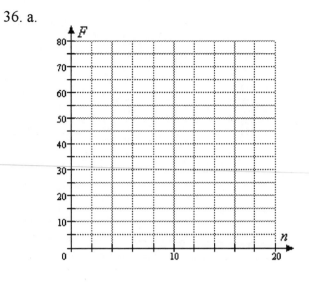

47.

x	x^2	$\log_{10} x$	$\log_{10} x^2$
1			
2			
3			
4			
5			
6			

48.

x	$\dfrac{1}{x}$	$\log_{10} x$	$\log_{10} \dfrac{1}{x}$
1			
2			
3			
4			
5			
6			

49.

x	$y = \log_e x$
1	0
2	0.693
4	
16	
$\dfrac{1}{2}$	
$\dfrac{1}{4}$	
$\dfrac{1}{16}$	

50.

x	$y = \log_f x$
1	0
2	0.431
4	
16	
$\dfrac{1}{2}$	
$\dfrac{1}{4}$	
$\dfrac{1}{16}$	

7.5 Properties of Logarithms

Exercise 1   Simplify $\log_b \dfrac{x}{y^2}$.

Exercise 2   Solve $\log_{10} x + \log_{10} 2 = 3$.

Rewrite the left side as a single logarithm.

Rewrite the equation in exponential form.

Solve for x.

Exercise 3   Solve $5(1.2)^{2.5x} = 77$.

Divide both sides by 5.

Take the log of both sides.

Apply Property (3) to simplify the left side.

Solve for x.

Exercise 4   Traffic on U.S. highways is growing by 2.7% per year. How long will it take the volume of traffic to double?

Exercise 5   Solve $N = N_0 \log_b (ks)$ for s.

Divide both sides by N_0.

Rewrite in exponential form.

Solve for k.

Homework 7.5

39a.

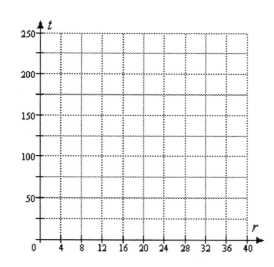

c.

Planet	Mercury	Venus	Earth	Mars	Jupiter	Saturn	Uranus	Neptune	Pluto
$X = \log r$									
$Y = \log t$									

d.

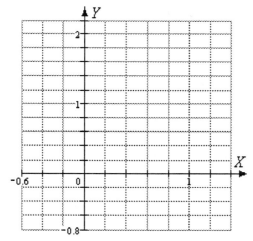

40. a.

Octave	0	1	2							
Frequency (hertz)	20	40	80							
Distance (mm)	35	31.5	28							

c.

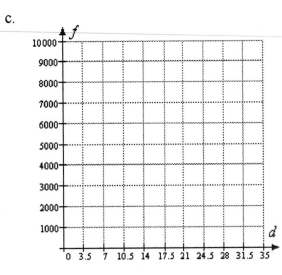

7.6 The Natural Base

Exercise 1   Use your calculator to evaluate.

    a. e^2 ***b.*** $e^{3.5}$ ***c.*** $e^{-0.5}$

    d. $\ln 100$ ***e.*** $\ln 0.01$ ***f.*** $\ln e^3$

Exercise 2   Solve $80 - 16e^{-0.2x} = 70.3$.

Subtract 80 from both sides and divide by -16.

Take the natural log of both sides.

Divide by -0.2.

Exercise 3   Solve $N = Ae^{-kt}$ for k.

Divide both sides by A.

Take the natural log of both sides.

Divide both sides by $-t$.

Exercise 4   From 1994 to 1998, the number of personal computers using the Internet grew according to the formula $N(t) = 2.8e^{0.85t}$, where $t = 0$ in 1994 and N is in millions.
  a. Evaluate $N(1)$. By what percent did the number of Internet users grow in one year?

  b. Express the growth law in the form $N(t) = N_0(1+r)^t$. (*Hint:* $e^k = 1 + r$.)

Exercise 5   A scientist isolates 25 grams of krypton-91, which decays according to the formula $N(t) = 25\,e^{-0.07t}$, where t is in seconds.
  a. Complete the table of values showing the amount of krypton-91 left at ten second intervals over the first minute.

t	0	10	20	30	40	50	60
$N(t)$							

  b. Use the table to choose a suitable window and graph the function $N(t)$.
  c. Write and solve an equation to answer the question: How long does it take for 60% of the krypton-91 to decay? (*Hint:* If 60% of the krypton-91 has decayed, 40% of the original 25 grams remains.)

Homework 7.6

1.

x	-10	-5	0	5	10	15	20
$f(x)$							

2.

x	-10	-5	0	5	10	15	20
$f(x)$							

3.

x	-10	-5	0	5	10	15	20
$f(x)$							

4.

x	-10	-5	0	5	10	15	20
$f(x)$							

17.

x	0	0.5	1	1.5	2	2.5
e^x						

18.

x	0	2	4	6	8	10
e^x						

19.

x	0	0.6931	1.3863	2.0794	2.7726	3.4657	4.1589
e^x							

20.

x	0	1.0986	2.1972	3.2958	4.3944	5.4931	6.5917
e^x							

35.

n	0.39	3.9	39	390
$\ln n$				

36.

n	0.64	6.4	64	640
$\ln n$				

37.

n	2	4	8	16
$\ln n$				

38.

n	5	25	125	625
$\ln n$				

49.

50.

Chapter 7 Review

1a.

t (years since 1974)	0	5	10	15	20
d (degrees awarded)					

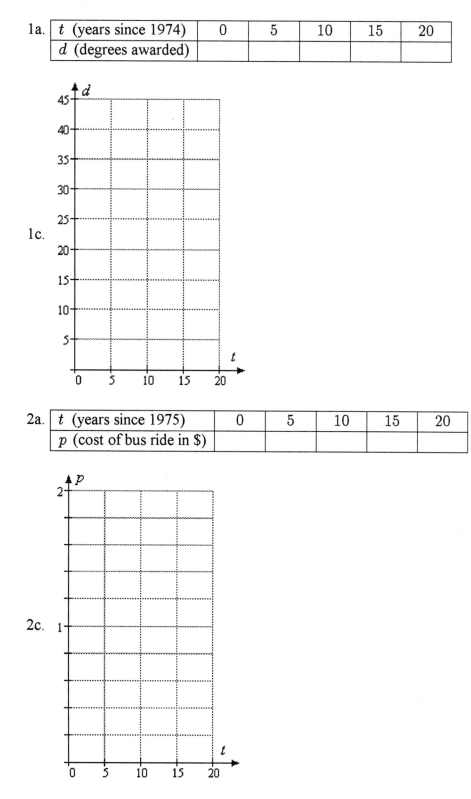

1c.

2a.

t (years since 1975)	0	5	10	15	20
p (cost of bus ride in \$)					

2c.

215

3a.

t (hours since 8 a.m.)	0	2	4	6	8	10
m (milligrams of medication)						

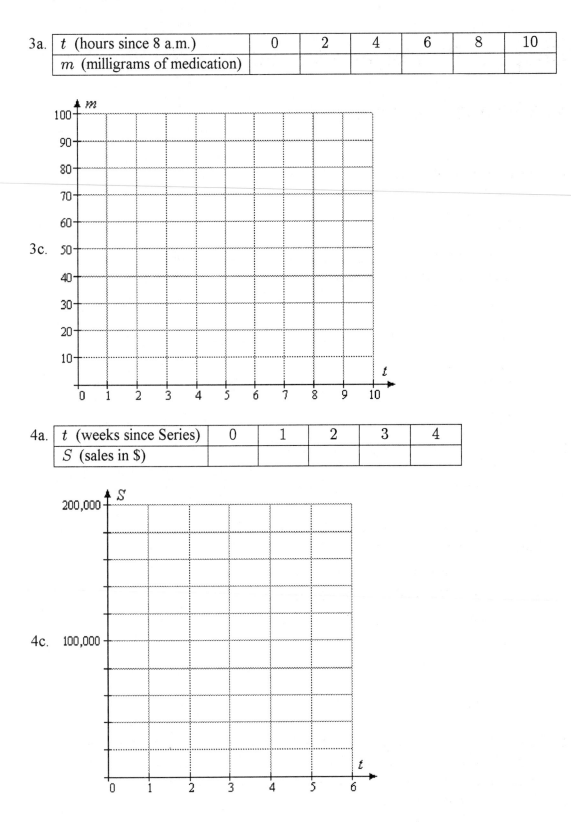

3c.

4a.

t (weeks since Series)	0	1	2	3	4
S (sales in $)					

4c.

Chapter 8   Polynomial and Rational Functions

Investigation 13   Polynomial Models

A.

  1. Complete Table 8.1 for the function P .

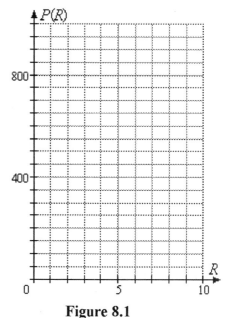

R	$P(R)$
0	
2	
4	
6	
8	
10	

Table 8.1

Figure 8.1

  2. Graph the function on the grid in Figure 8.1.

B.

  1. How much material will be needed for each cubicle if it measures 10 feet by 10 feet?
  Step 1   Compute the areas:

     Area of ceiling   = _____

     Area of each wall   = _____

     Area of door   = _____

  Step 2

              Area of ceiling + 4(Area of each wall) − Area of door

     Total =

  2. Now follow the same steps for a cubicle of arbitrary size. Let d represent the dimensions (length and width) of the cubicle, and write a function for the amount of sound-proofing material A needed for each cubicle in terms of d .
  Step 1   Compute the areas:

     Area of ceiling   = _____

     Area of each wall   = _____

     Area of door   = _____

217

Step 2

$$\text{Area of ceiling} + 4(\text{Area of each wall}) - \text{Area of door}$$

$$A(d) =$$

3. Complete Table 8.2 for the function $A(d)$.

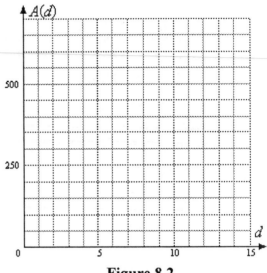

d	$A(d)$
5	
6	
8	
12	
15	

Table 8.2

Figure 8.2

C.

x	$A(x)$
0.5	
1.0	
1.5	
2.0	
2.5	

Table 8.3

8.1 Polynomial Functions

Exercise 1 Multiply $(y+2)(y^2 - 2y + 3)$

Exercise 2 Write $(5 + x^2)^3$ as a polynomial.

Exercise 3 Factor $125n^3 - p^3$.

8.2 Graphing Polynomial Functions

Exercise 1a. Complete the table of values for $C(x) = -x^3 - 2x^2 + 4x + 4$.

x	-4	-3	-2	-1	0	1	2	3	4
y									

b. Graph $y = C(x)$ in the standard window. Compare the graph to the graphs in Example 1: What similarities do you notice? What differences?

Exercise 2a. Complete the table of values for $Q(x) = -x^4 - x^3 - 6x^2 + 2$.

x	-4	-3	-2	-1	0	1	2	3	4
y									

b. Graph $y = Q(x)$ in the window

$$\text{Xmin} = -5 \quad \text{Xmax} = 5$$
$$\text{Ymin} = -15 \quad \text{Ymax} = 10$$

Compare the graph to the graphs in Example 2: What similarities do you notice? What differences?

Exercise 3a. Find the zeros of

$$P(x) = -x^4 + x^3 + 2x^2$$

by factoring.

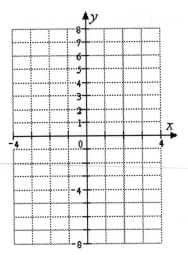

b. Sketch a rough graph of $y = P(x)$
by hand.

Exercise 4   Sketch a rough graph of

$$f(x) = (x+3)(x-1)^2$$

by hand. Label the x- and y-intercepts.

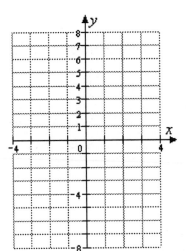

Homework 8.2

19.

20.

21.

22.

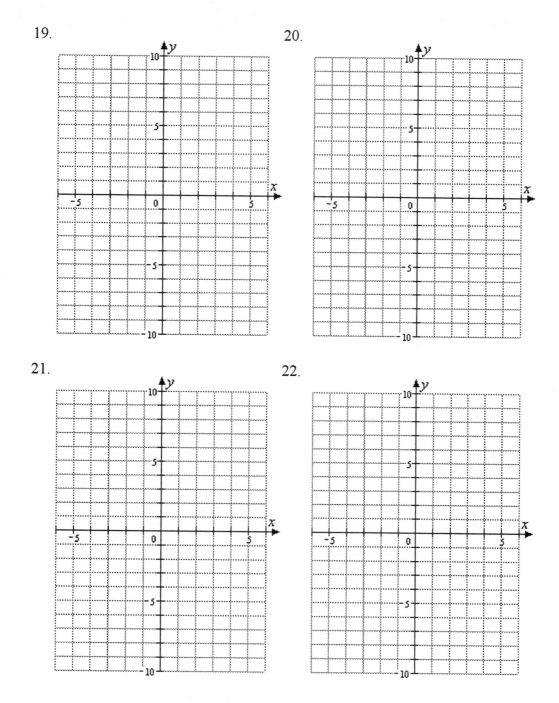

23.

24.

25.

26.

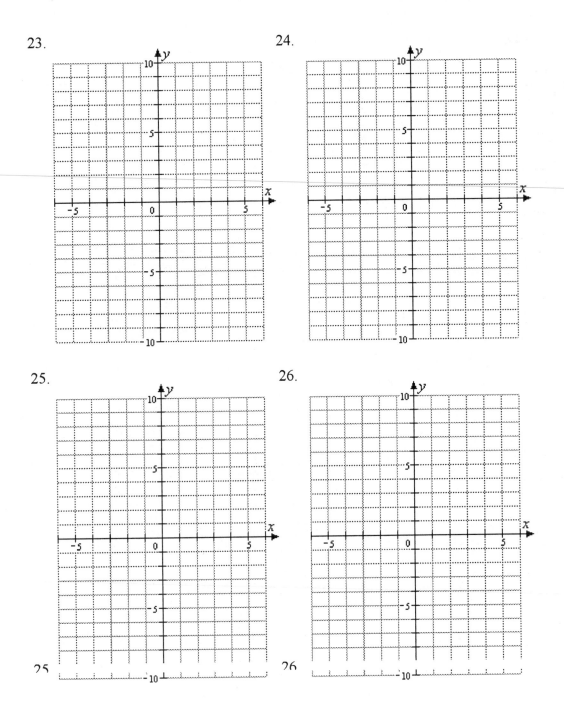

35.

36.

37.

38.

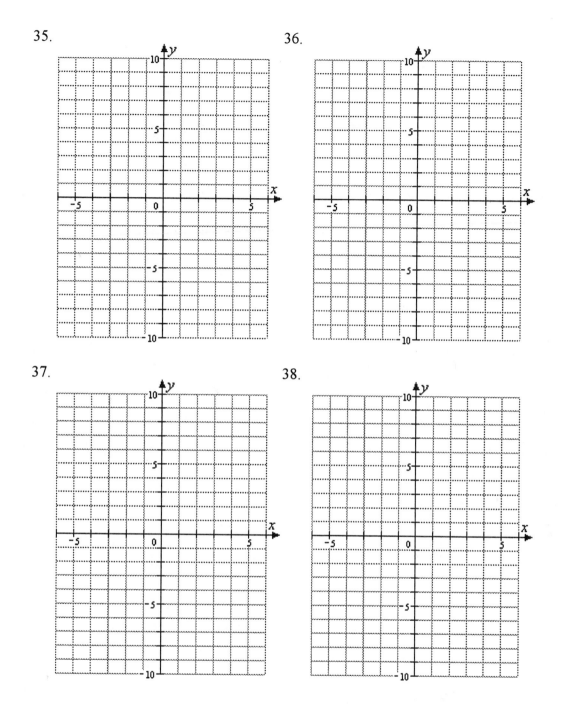

39.

40.

41.

42.

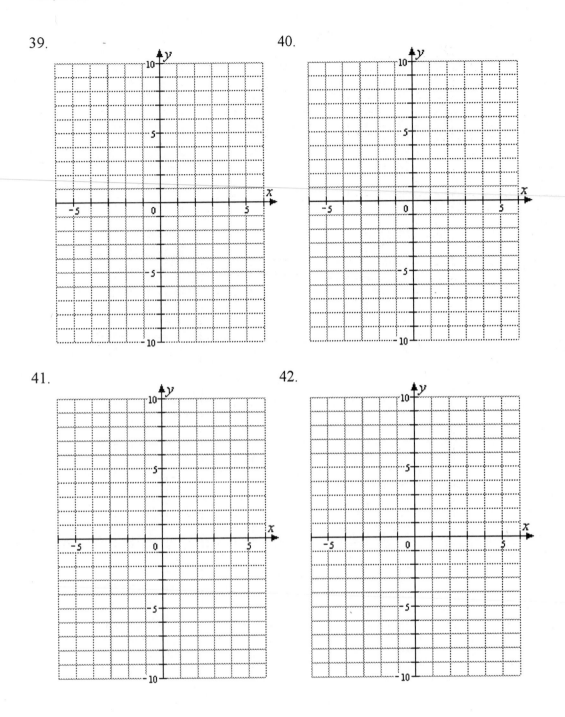

43.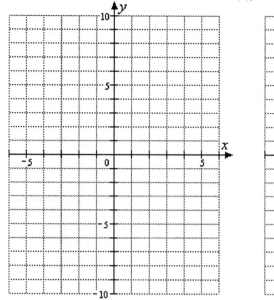

44.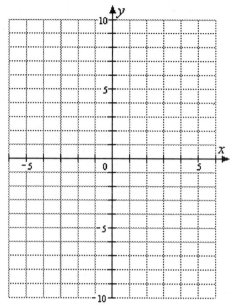

8.3   Rational Functions

Exercise 1   Find the domain of $F(x) = \dfrac{x-2}{x+4}$.

Exercise 2   Find the vertical asymptotes of $G(x) = \dfrac{4x^2}{x^2-4}$.

Exercise 3   Find the horizontal asymptote of $G(x) = \dfrac{4x^2}{x^2-4}$.

Exercise 4   Locate the horizontal and vertical asymptotes and sketch the graph of

$$k(x) = \dfrac{x^2-1}{x^2-9}.$$

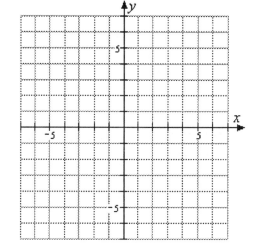

Vertical asymptotes:

Horizontal asymptote:

x-intercepts:

y-intercept:

x	-4	-1	1	4
$k(x)$				

Homework 8.3

1.

v	0	5	10	15	20	25	30	35	40	45	50
t											

2.

v	0	4	8	12	16	20	24	28	32	36
t										

3.

p	0	15	25	40	50	75	80	90	100
C									

4.

p	0	15	25	40	50	75	80	90	100
C									

5.

n	100	200	400	500	1000	2000	4000	5000	8000
C									

6.

p	0.25	0.50	1.00	1.25	1.50	1.75	2.00	2.25	2.50	2.75	3.00
Demand											
Revenue											

7.

x	10	20	30	40	50	60	70	80	90	100
C										

8.

x	10	20	30	40	50	60	70	80	90	100
C										

9.

x	1	2	3	4	5	6	7
h							
V							

10.

x	1	2	3	4	5	6	7	8
h								
S								

11.

v	-100	-75	-50	-25	0	25	50	75	100
P									

12.

v	100	200	300	400	500	600	700	800	900	1000
h										

13.

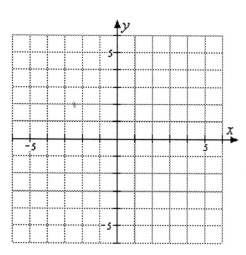

14.

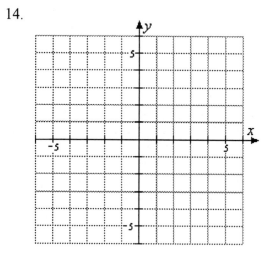

15.

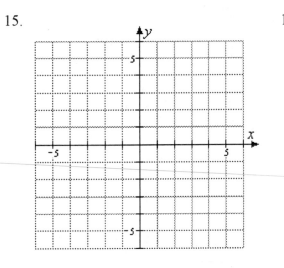

16.

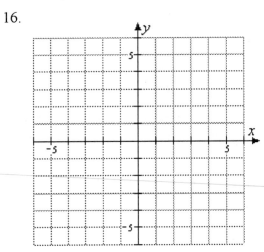

17.

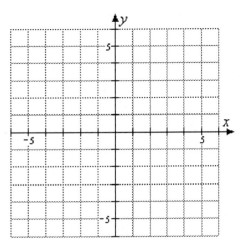

18.

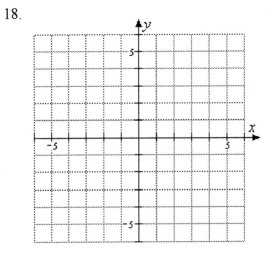

19.

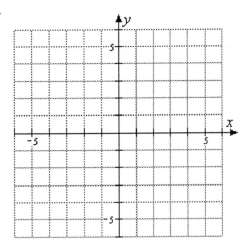

20.

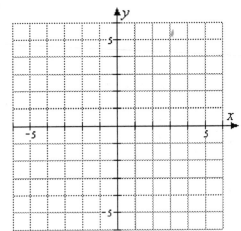

21.

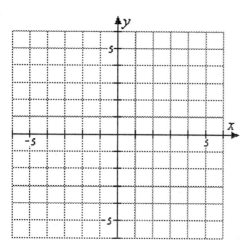

22.

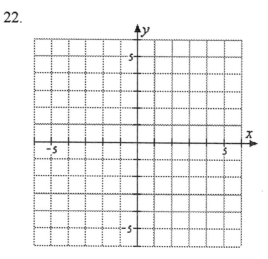

23.

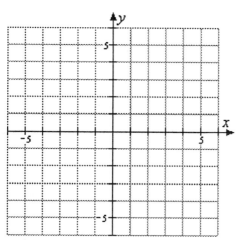

24.

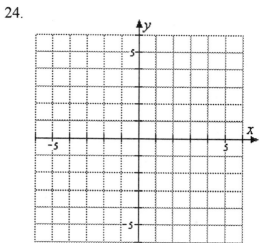

25.

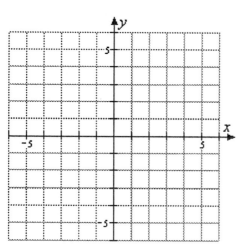

26.

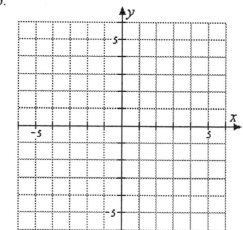

27.

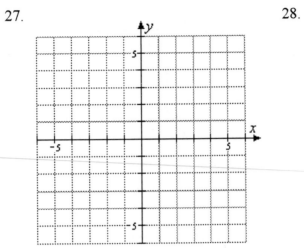

28.

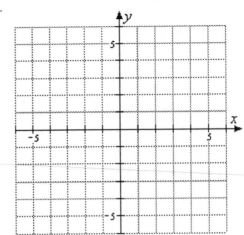

29.

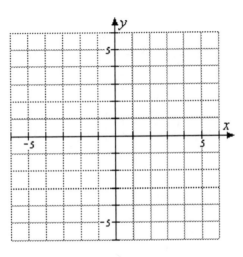

30.

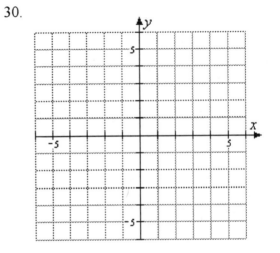

31.

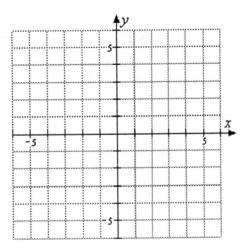

32.

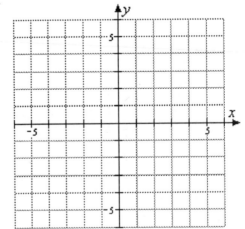

37.

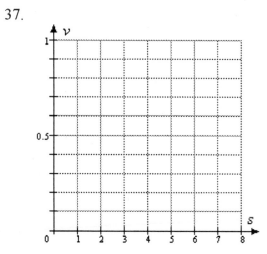

38.

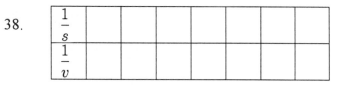

$\dfrac{1}{s}$						
$\dfrac{1}{v}$						

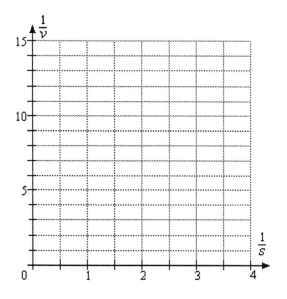

39.

s							
$\dfrac{s}{v}$							

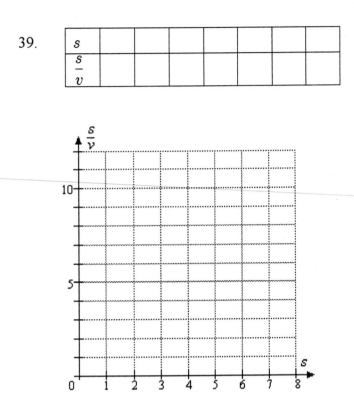

Midchapter Review

9.

10.

11.

12.

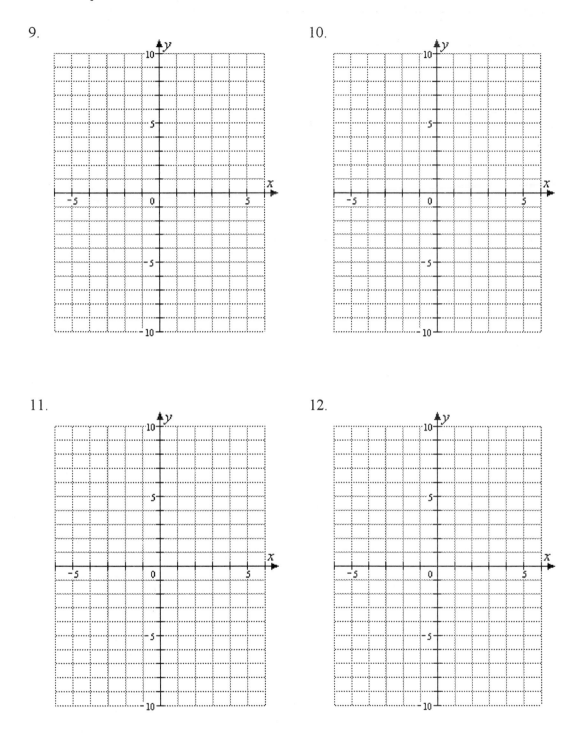

15.

16.

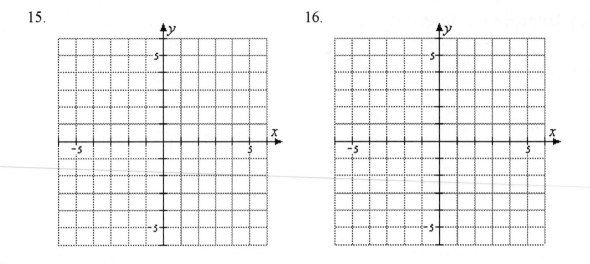

8.4 Operations on Algebraic Fractions

Exercise 1   Reduce $\dfrac{2a + 8b}{4b}$.

Exercise 2   Reduce $\dfrac{16t^2 - 4}{4t + 4}$.

Exercise 3   Reduce $\dfrac{m^2 - 9}{m^2 - 3m - 18}$.

Exercise 4   Find the product $\dfrac{15n^2 - 5n}{3p^2} \cdot \dfrac{6p^3}{18n^3 + 2n}$.

Exercise 5   Find the quotient $\dfrac{x^2 - 4y^2}{4y^2} \div \dfrac{x^2 - 3xy + 2y^2}{3xy}$.

Exercise 6   Divide $\dfrac{6a^3 + 2a^2 - a}{2a^2}$.

Exercise 7   Divide $\dfrac{4 + 8y^2 - 3y^3}{3y + 1}$.

Homework 8.4

65.

66.

67.

68.

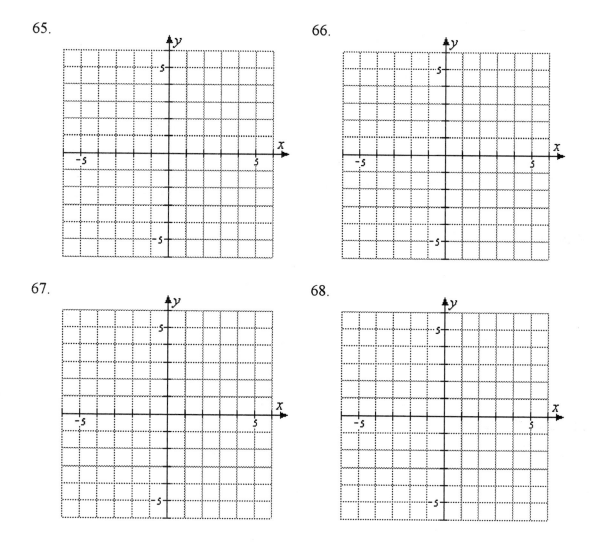

8.5 More Operations on Algebraic Fractions

Exercise 1   Subtract $\dfrac{a+1}{a^2-a} - \dfrac{5-3a}{a^2-a}.$

Exercise 2   Find the LCD for the fractions $\dfrac{1}{2x^3(x-1)^2}$ and $\dfrac{3}{4x^2-4x}.$

Exercise 3   Add $\dfrac{2}{x^2-x-2} + \dfrac{2}{x^2+2x+1}.$

Exercise 4   Simplify $\dfrac{x-2}{x-\dfrac{4}{x}}.$

Exercise 5   Simplify $(1+x^{-1})^{-1}.$

8.6 Equations that Include Algebraic Fractions

Exercise 1 Solve $\dfrac{x^2}{x+4} = 2$.

Exercise 2 Solve $\dfrac{x}{6-x} = \dfrac{1}{2}$.

Exercise 3   Solve $\dfrac{9}{x^2 + x - 2} + \dfrac{1}{x^2 - x} = \dfrac{4}{x - 1}.$

Exercise 4   Solve for a: $\dfrac{2ab}{a + b} = H.$

Chapter 8 Review

9.

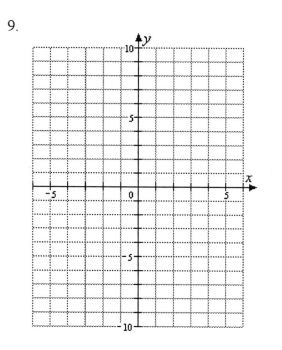

10.

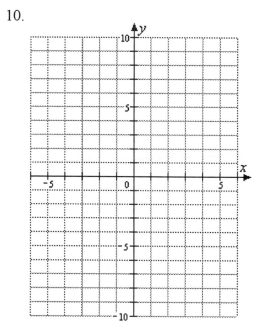

11.

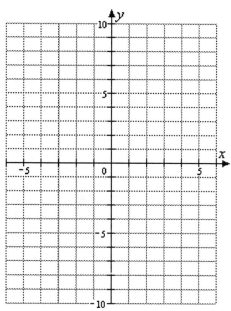

12.

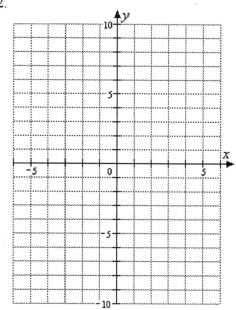

13.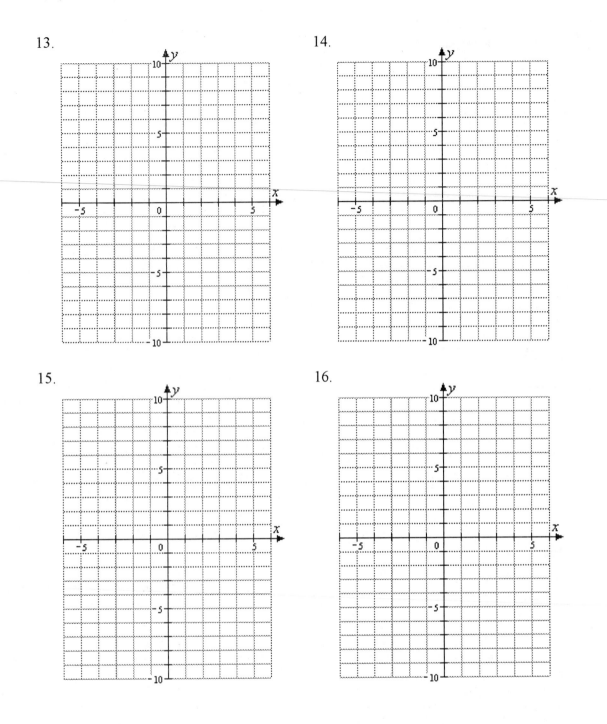

14.

15.

16.

17.

18.

25.

26.

27.

28.

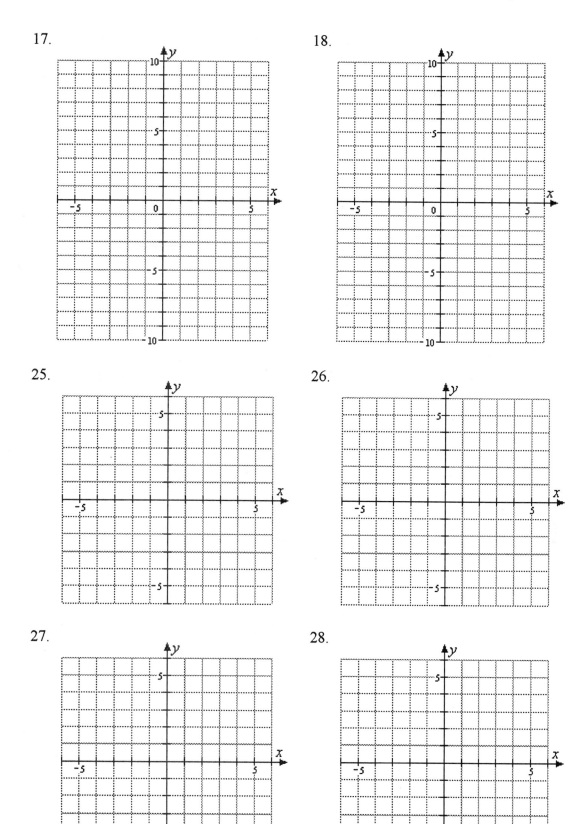

29.

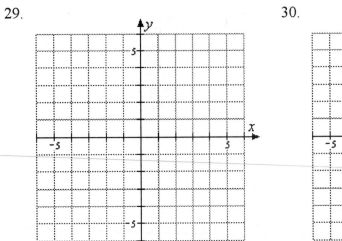

30.

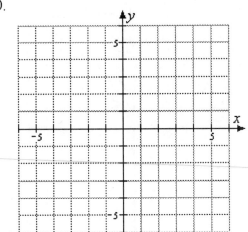

Chapter 9 Sequences and Series

Investigation 14   Scheduling a Soccer League

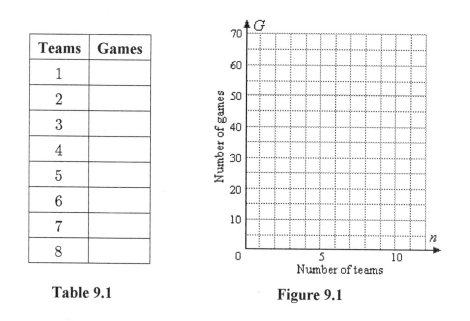

Teams	Games
1	
2	
3	
4	
5	
6	
7	
8	

Table 9.1

Figure 9.1

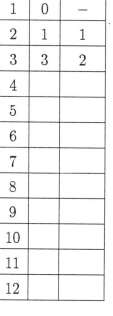

n	G	ΔG
1	0	–
2	1	1
3	3	2
4		
5		
6		
7		
8		
9		
10		
11		
12		

Table 9.2

9.1 Sequences

Exercise 1   Let A_n be the number of dollars in a bank account at the end of the nth year since the account began. Write an equation using subscript notation for each of the following statements.

    a. The account had $1149.90 at the end of the second year.

    b. At the end of year 4 the account had $85.75 more than at the end of year 3.

    c. At the end of year 5 the account had 1.07 times as much as at the end of year 4.

Exercise 2   Find the first four terms in the sequence with the general term
$b_n = (n + 1)^2 - n^2$.

Exercise 3   You just finished your last cup of coffee and have 100 mg of caffeine in your system. For each hour that passes, the amount of caffeine in your system decreases by 14%. How much caffeine is in your system at the start of each of the next four hours?

Exercise 4   The financing agreement on your $12,000 car requires you to pay $1200 now and $208.87 a month for five years. What is the total amount you have paid after each of the first 4 payments?

Exercise 5   Find a recursive definition for the sequence in Exercise 3 if you do not drink any more coffee for n hours.

Homework 9.1

53.

n	A_n	B_n	C_n	S_n
0	2	0	0	2
1	2	2	0	6
2	4	2	2	14
3				
4				
5				
6				
7				
8				

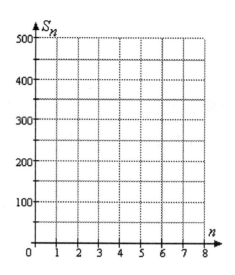

54.

n	S_n	G_n
0		
1		
2		
3		
4		
5		
6		
7		
8		
9		
10		
11		
12		
13		
14		
15		
16		
17		
18		
19		
20		

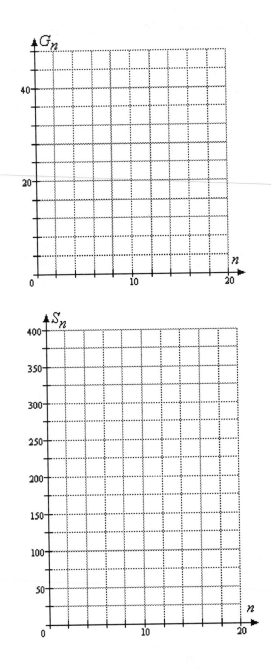

9.2 Arithmetic and Geometric Sequences

Exercise 1   Find the first four terms an arithmetic sequence with first term 100 and common difference -30.

Exercise 2   Find the hundredth term of the arithmetic sequence $1, 3, 5, 7, \ldots$.

Exercise 3   Find the first four terms of a geometric sequence whose first term is 1000 and whose common ratio is 1.06.

Exercise 4   Identify the following sequences as arithmetic, geometric, or neither.

    a. $1, 4, 9, 16, \ldots$ **b.** $2, 6, 18, 54, \ldots$ **c.** $\dfrac{1}{3}, \dfrac{1}{2}, \dfrac{2}{3}, \dfrac{5}{6}, \ldots$

Exercise 5 Find the twelfth term of the geometric sequence $\dfrac{3}{32}, \dfrac{3}{16}, \dfrac{3}{8}, \dfrac{3}{4}, \ldots$

Exercise 6 Find a non-recursive definition for each sequence.

a. $c_1 = 100,$ $c_{k+1} = 1.03c_k$ **b.** $d_1 = 2,$ $d_k = d_{k-1} - 5$

Homework 9.2

49.

50.

51.

52.

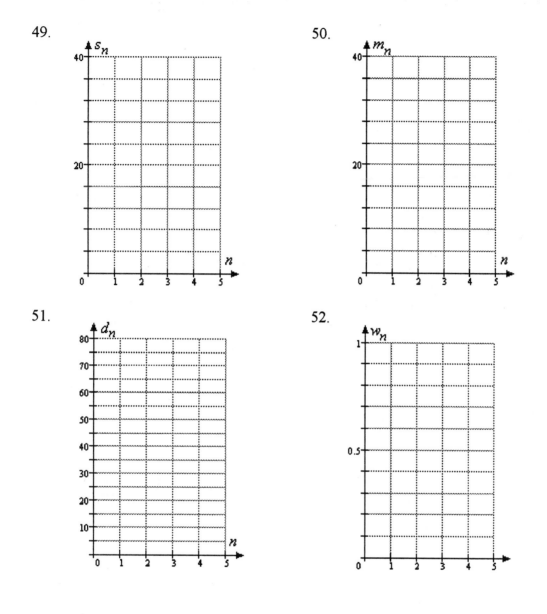

53.

54.

55.

56.

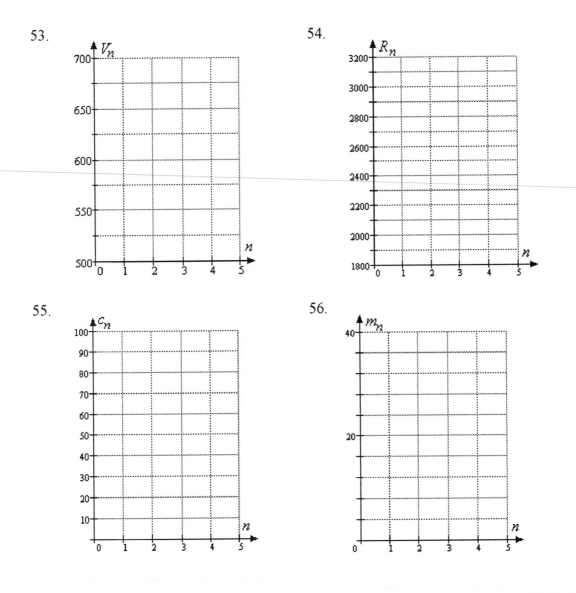

9.3 Series

Exercise 1 Find the series S_5 for the sequence with general term $b_n = n^2$.

Exercise 2 Find the sum of the first 100 terms in the arithmetic sequence $2, 5, 8, 11, \ldots$.

Exercise 3 Find the sum of the first 10 terms in the geometric sequence $100, 110, 121, \ldots$.

9.4 Sigma Notation and Infinite Geometric Series

Exercise 1   Use sigma notation to represent the sum of the first 50 terms of the sequence

$$5, 8, 11\ldots, 3k + 2, \ldots$$

Exercise 2   Compute the value of each series.

a. $\displaystyle\sum_{k=1}^{50} (3k + 2)$

b. $\displaystyle\sum_{n=1}^{8} 10^n$

Exercise 3   Compute the value of each series.

a. $\displaystyle\sum_{k=0}^{20} \frac{1}{3}$

b. $\displaystyle\sum_{k=0}^{4} \frac{k}{k+1}$

Exercise 4   Find the sum, if it exists, or state that the series does not have a sum.

a. $\displaystyle\sum_{j=0}^{\infty} 13\left(\frac{7}{6}\right)^{j}$

b. $\displaystyle\sum_{m=0}^{\infty} 5.9\,(0.9)^{m}$

Exercise 5   Find a fraction equivalent to $0.\overline{8}$.

9.5 The Binomial Expansion

Investigation 15   Powers of Binomials

Expand each power and fill in the blanks. Arrange the terms in each expansion in descending powers of a.

1. $(a+b)^0 = $ _____.

2. $(a+b)^1 = $ _____.

3. $(a+b)^2 = $ _____.

4. $(a+b)^3 = $ _____.

5. $(a+b)^4 = $ _____.

6. $(a+b)^5 = $ _____.

7.   Do you see a relationship between the exponent n and the number of terms in the expansion of $(a+b)^n$? (Notice that for $n=0$ we have $(a+b)^0 = 1$,   which has one term.) Fill in Table 9.7.

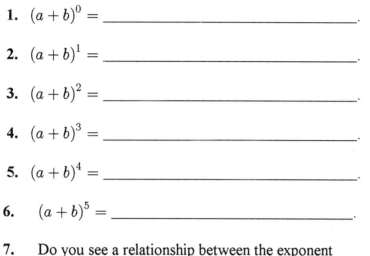

n	Number of terms in $(a+b)^n$
0	
1	
2	
3	
4	
5	

Table 9.7

8. First observation: In general, the expansion of $(a+b)^n$ has _____ terms.

9. Next we'll consider the exponents on a and b in each term of the expansions. Refer to your expanded powers in parts 1–5, and fill in Table 9.8.

Table 9.8

n	First term of $(a+b)^n$	Last term of $(a+b)^n$	Sum of exponents on a and b in each term
0			
1			
2			
3			
4			
5			

10. **Second observation:** In any term of the expansion of $(a+b)^n$, the sum of the exponents on a and on b is _____.

11. Complete the following tables for the cases $n = 3$, $n = 4$, and $n = 5$.

Case $n = 3$:

k	0	1	2	3
Exponent on a	3			
Exponent on b	0			

Table 9.10

Case $n = 4$:

k	0	1	2	3	4
Exponent on a	4				
Exponent on b	0				

Table 9.11

Case $n = 5$:

k	0	1	2	3	4	5
Exponent on a	5					
Exponent on b	0					

Table 9.12

12. **Third observation:** The variable factors of the kth term in the expansion of $(a+b)^n$ may be expressed as _____ . (Fill in the correct powers in terms of n and k for a and b.)

Exercise 1   Expand $(r + 2s)^3$.

Exercise 2   Evaluate the binomial coefficients.

    a. $_5C_4$ ***b.*** $_5C_0$

Investigation 16   Pascal's Triangle

To get a clearer picture of the binomial coefficients, consider again the expansions of $(a + b)^n$ you calculated in Investigation 15, but this time look only at the numerical coefficients of each term:

$$
\begin{array}{llccccccccccc}
n = 0 & & & & & & & 1 & & & & & \\
n = 1 & & & & & & 1 & & 1 & & & & \\
n = 2 & & & & & 1 & & 2 & & 1 & & & \\
n = 3 & & & & 1 & & 3 & & 3 & & 1 & & \\
n = 4 & & & 1 & & 4 & & 6 & & 4 & & 1 & \\
n = 5 & & 1 & & 5 & & 10 & & 10 & & 5 & & 1
\end{array}
$$

1. Each row of Pascal's triangle begins with the number _____ and ends with the number _____ .

2. The second number and the next-to-last number in the nth row are _____ .

3. Starting with the row $n = 2$, any number in the triangle (except the first and last 1s in each row) can be found by _____ .

4. Using your answer to Step 3, continue Pascal's triangle to include the row for $n = 6$.

5. Use Pascal's triangle to find the binomial coefficient $_6C_4$.

6. Expand: $(a+b)^6 =$

7. Expand: $(x-2)^6 =$ _____

8. Continue Pascal's triangle to include the row for $n=7$.

Exercise 3   Write the expanded form of $(w+10z)^6$.

Exercise 4   Write $\dfrac{5!}{3!2!}$ in expanded form and simplify.

Exercise 5   Evaluate the binomial coefficient $_8C_6$.

Exercise 6   Find the coefficient of s^3t^4 in the expansion of $(s+t)^7$.

Exercise 7a. Use sigma notation to write the expanded form for $(2m + 3n)^7$.

b. Find the term containing m^3, and simplify.

Investigation 17   Other Properties of Pascal's Triangle

Pascal's triangle has other interesting properties besides providing the binomial coefficients.

1. Add up the numbers in each row of your Pascal's triangle and fill in Table 9.13.

b. Find the term containing m^3, and simplify.

Now add entries of Pascal's triangle along the diagonal lines indicated below. Fill in two more row of Pascal's triangle and find two more diagonal sums.

Row n	Sum of Entries
0	
1	

4	
5	
6	

Table 9.13

3. Write the sequence of numbers obtained by taking sums along the indicated diagonal lines.

_____, _____, _____, _____, _____, _____, _____, $\cdots$

4. What is the name of this sequence? (*Hint*: See Problem 45 of Section 9.1.)

5. Find the number of different choices you can make for a four-flavor ice cream sundae if there are six flavors to choose from. (*Caution*: When you are deciding which entry in the row corresponds to $k = 4$, remember that the leftmost entry corresponds to $k = 0$.)

6. How many ways can you choose a debate team of three members from a debate club of five members?

7. Evaluate the series $\displaystyle\sum_{r=0}^{10} {}_{10}C_r$

8. Evaluate the series $\displaystyle\sum_{r=0}^{12} {}_{12}C_r$

9. How many ways can you choose 6 objects from a set of 49? (This tells the total number of possible choices in a lottery that requires a participant to pick 6 numbers from 1 to 49.)

10. How many ways can you choose 5 cards out of a deck of 52 distinct cards? (This tells the total number of different poker hands.)

Chapter 10 More About Functions

Investigation 18 Transformations of Graphs

A. How is the graph of a function affected if we add a constant to the formula? In other words, how is the graph of $y = f(x) + k$ different from the graph of $y = f(x)$?

B. How is the graph of a function affected if we add a constant to the independent variable? In other words, how is the graph of $y = f(x + h)$ different from the graph of $y = f(x)$?

C. How is the graph of a function affected if we multiply the formula by a constant? In other words, how is the graph of $y = af(x)$ different from the graph of $y = f(x)$?

10.1 Transformations of Functions

Exercise 1   **a.** Graph the function $f(x) = |x| + 1$.
   b. How is the graph different from the graph of $y = |x|$?

Exercise 2   **a.** Graph the function $f(x) = |x + 1|$.
   b. How is the graph different from the graph of $y = |x|$?

Exercise 3   **a.** Graph the function $f(x) = |x - 2| - 1$.
   b. How is the graph different from the graph of $y = |x|$?

Exercise 4   **a.** Graph the function $f(x) = 2|x|$.
   b. How is the graph different from the graph of $y = |x|$?

Homework 10.1

1.

2.

3.

4.

5.

6.

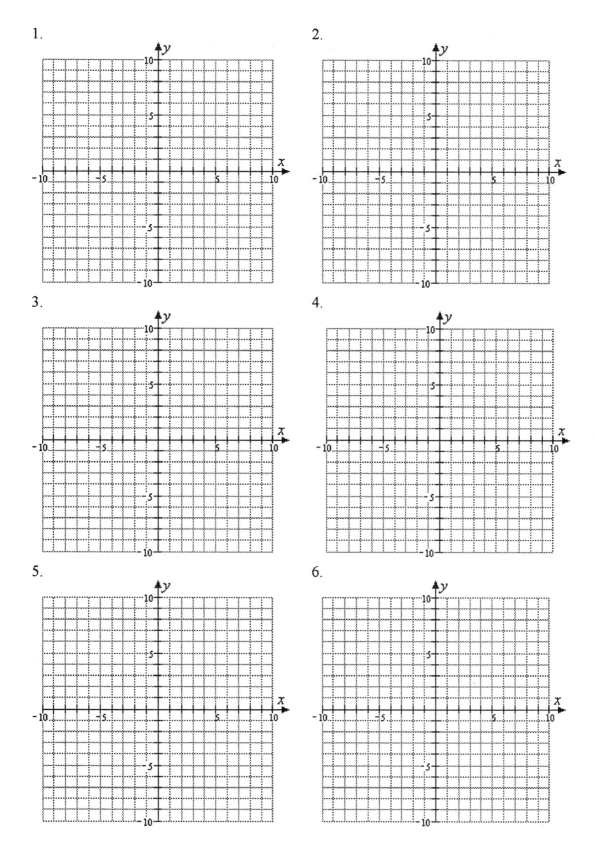

7.

8.

9.

10.

11.

12.

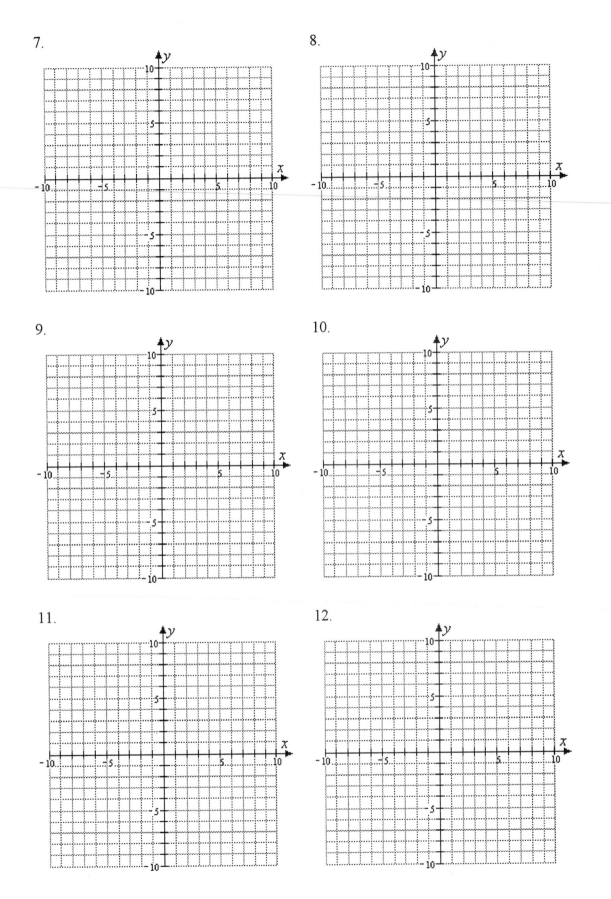

13.

14.

15.

16.

17.

18.

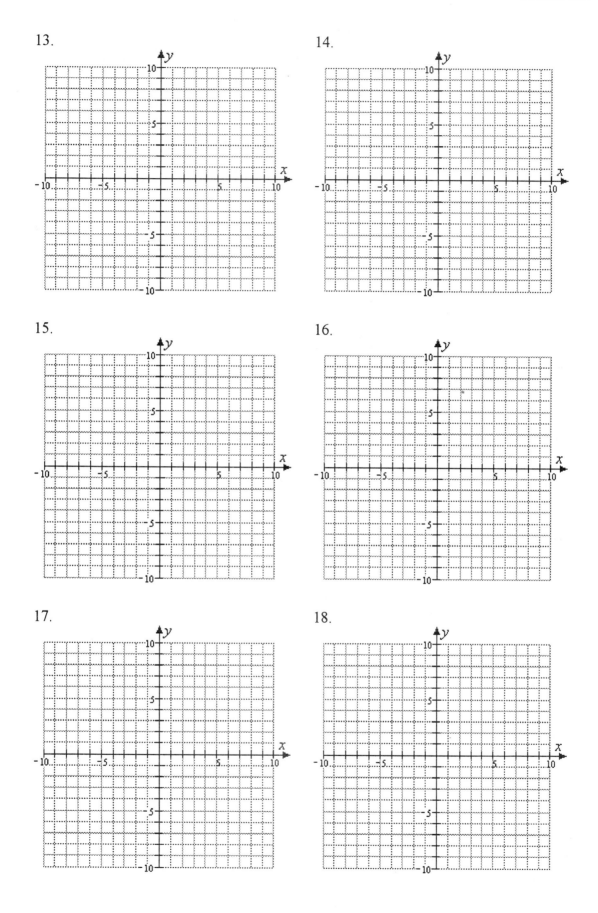

19.

20.

21.

22.

23.

24.

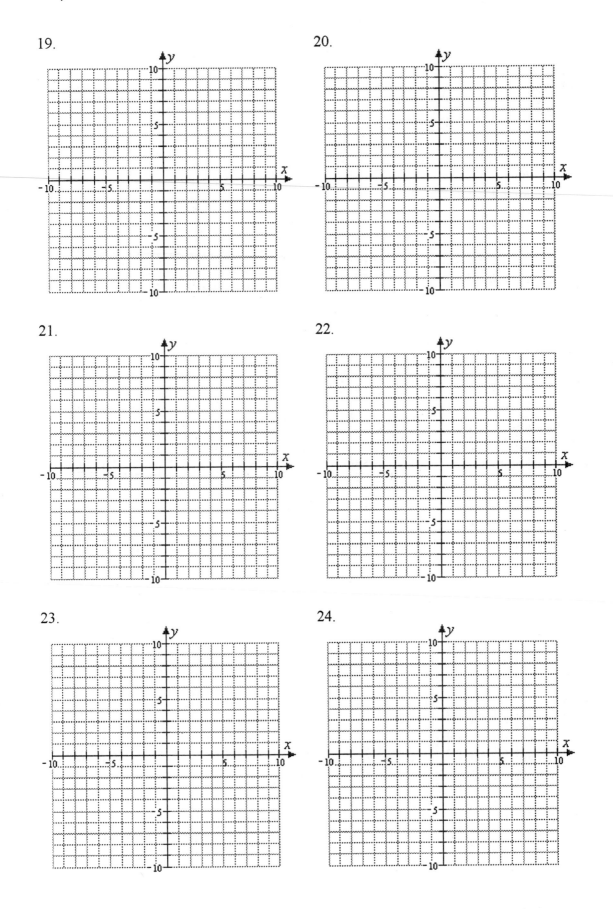

25.

26.

27.

28.

29.

30.

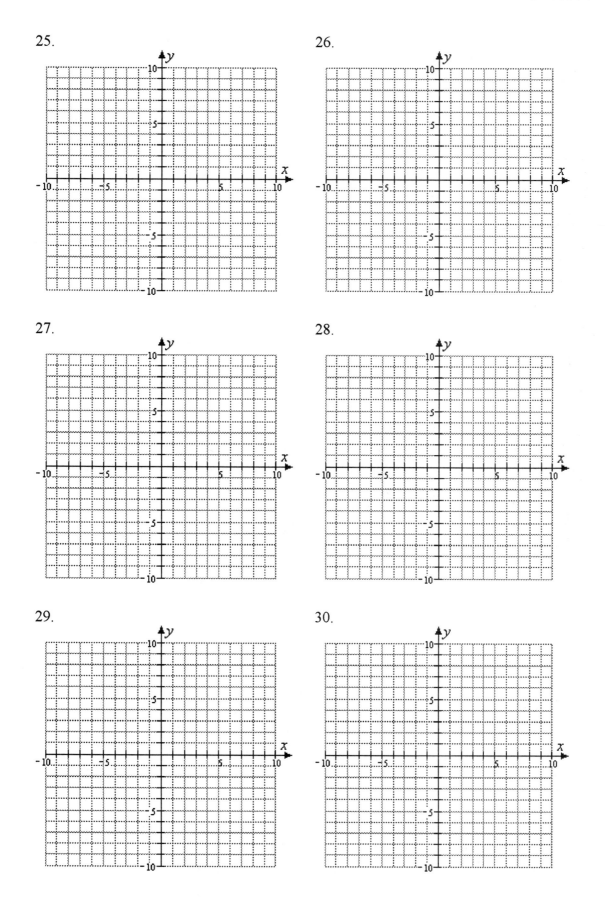

31.

32.

33.

34.

45.

46.

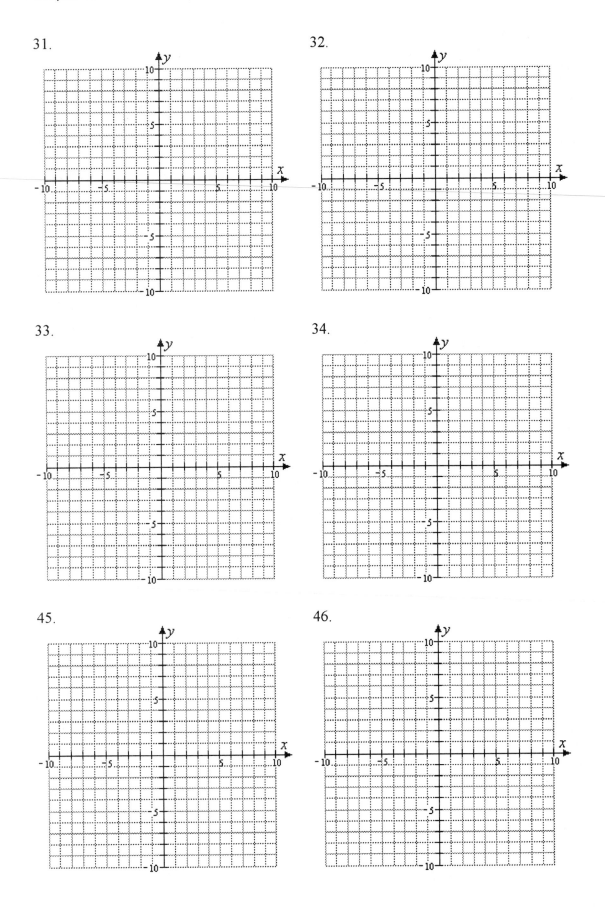

47.

48.

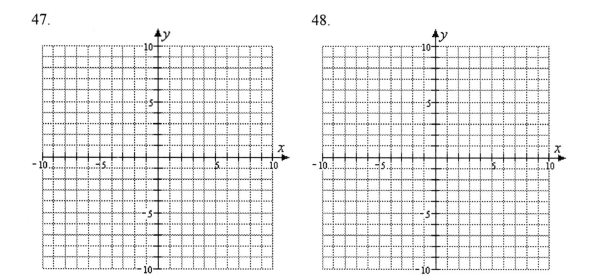

10.2 Using Function Notation

Exercise 1 Find $F(-1)$ if $F(x) = -x^2 - x + 2$.

Exercise 2 Evaluate the function $G(t) = 1 - t^2$ for the following expressions.

a. $t = 3w$

b. $t = s + 1$

Exercise 3 Let $F(r) = \dfrac{3}{r} + 1$, and evaluate each expression.

a. $F(-1) + F(3)$

b. $F(-1 + 3)$

c. $-1F(r) + 3$

Exercise 4 Graph the piecewise defined function

$$g(x) = \begin{cases} -1 & \text{if } x \leq -3 \\ x + 2 & \text{if } x > -3 \end{cases}$$

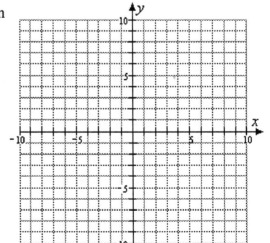

281

Homework 10.2

21.

22.

23.

24.

25.

26.

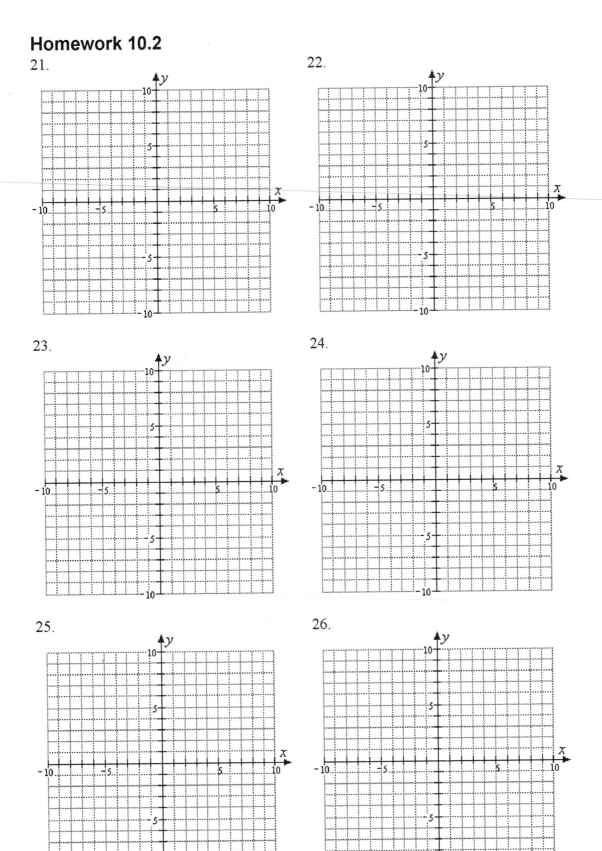

27.

28.

29.

30.

31.

32.

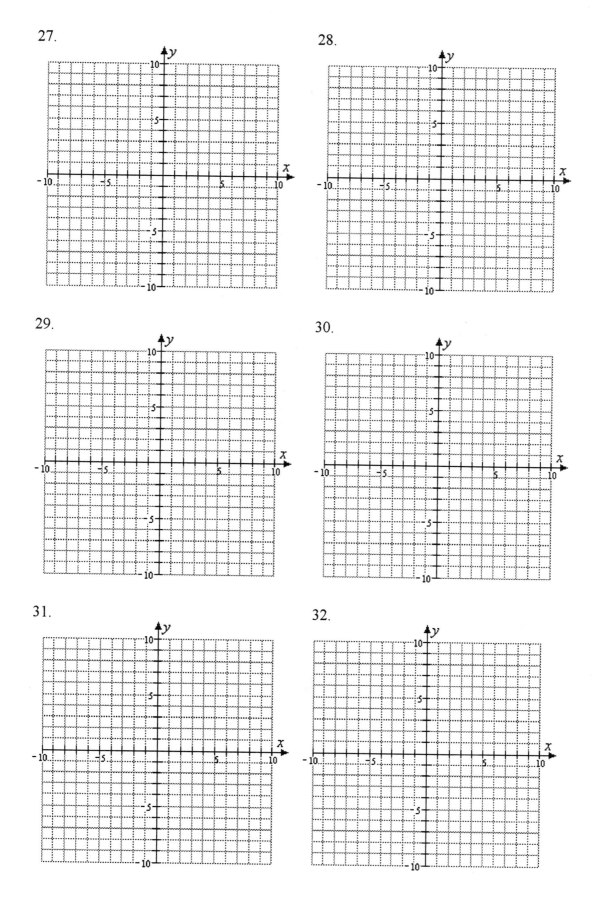

33.

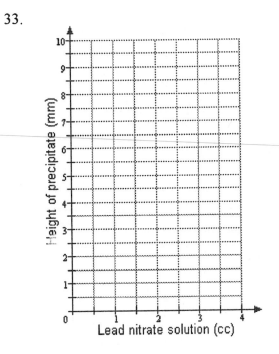

35.

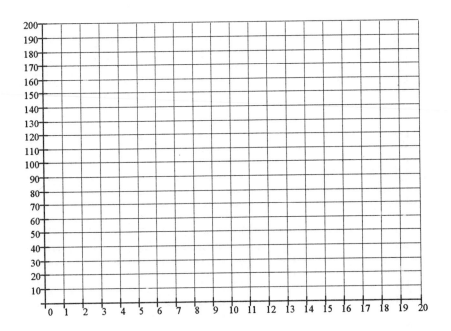

36.

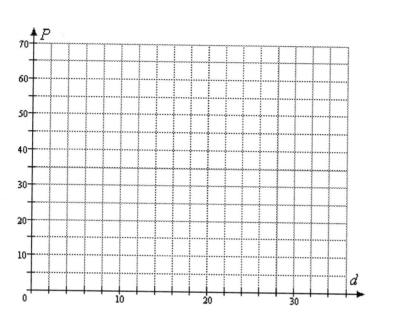

10.3 Inverse Functions

Exercise 1   Suppose g is the inverse function for f, and we know the following function values for f:

$$f(-1) = 0, \quad f(0) = 1, \quad f(1) = 2.$$

Find $g(0)$ and $g(1)$.

Exercise 2   Find a formula for $g(x)$, the inverse of the function $f(x) = \frac{x}{3} - 5$

Exercise 3a.   Find $g(x)$, the inverse of the function $f(x) = \sqrt[5]{x+2}$.

b.   Show that g "undoes" the effect of f on $x = -3$.

c.   Show that f "undoes" the effect of g on $x = 2$.

Exercise 4   Graph the function

$$f(x) = \sqrt[5]{x} + 2$$

and its inverse (which you found in Exercise 3) on the same set of axes.

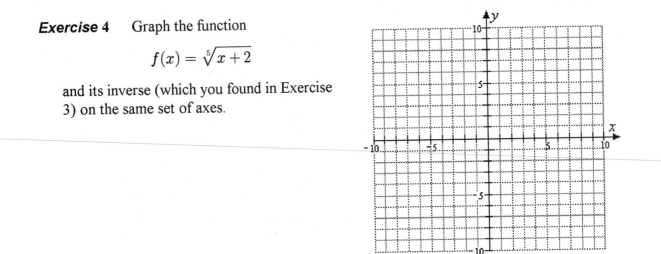

Exercise 5   Which of the functions whose graphs are shown below have inverses that are also functions?

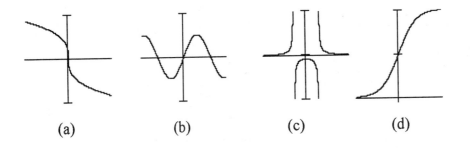

(a)   (b)   (c)   (d)

Exercise 6   If $f(x) = \dfrac{1}{x+1}$, find $f^{-1}(1)$.

Homework 10.3

1.

2.

3.

4.

5.

6.

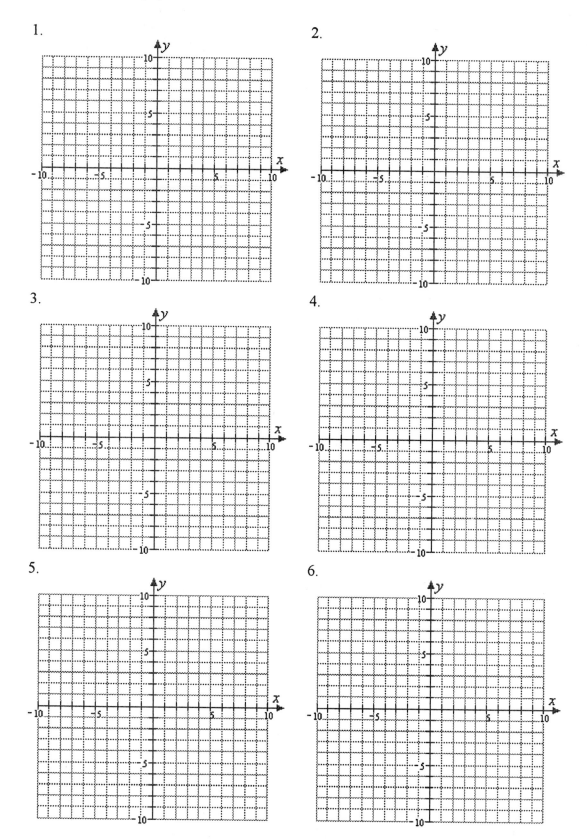

7.

8.

9.

10.

11.

12.

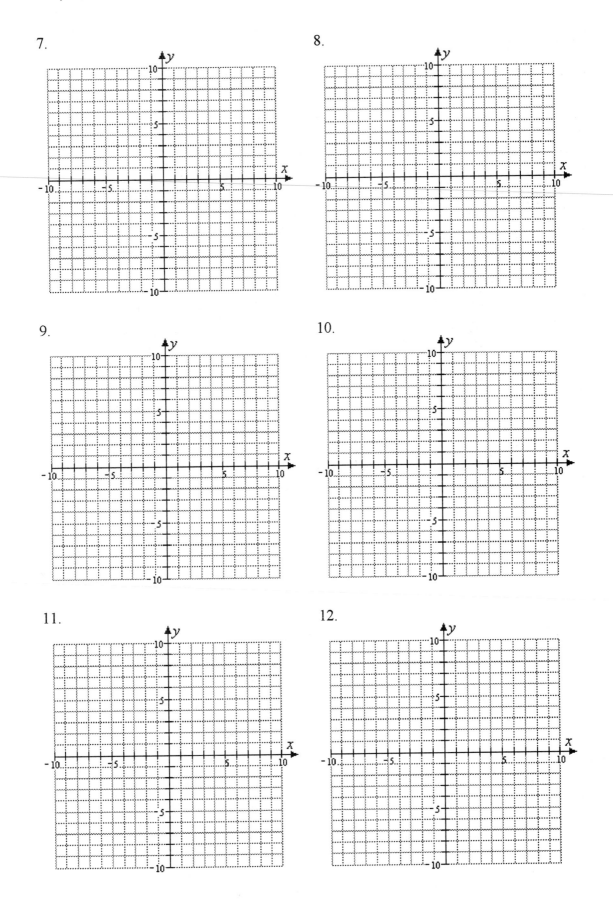

13.

14.

15.

16.

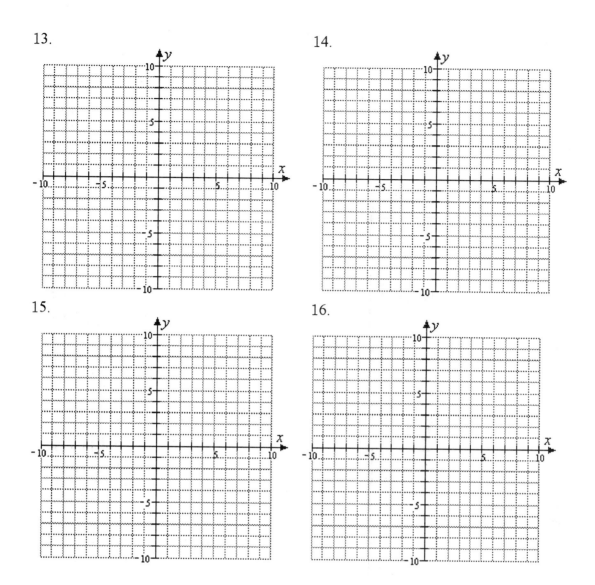

Midchapter Review

1.

2.

3.

4.

5.

6.

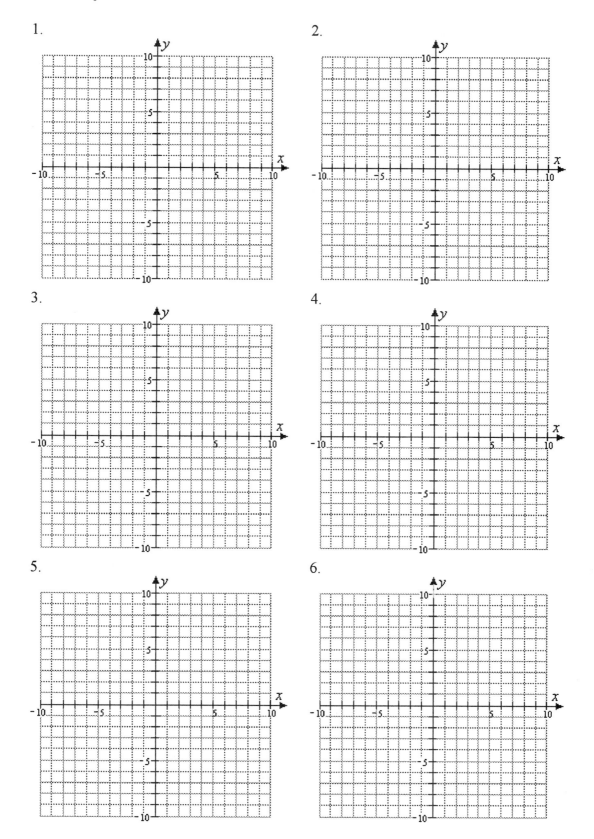

7.

8.

15.

16.

23.

24.

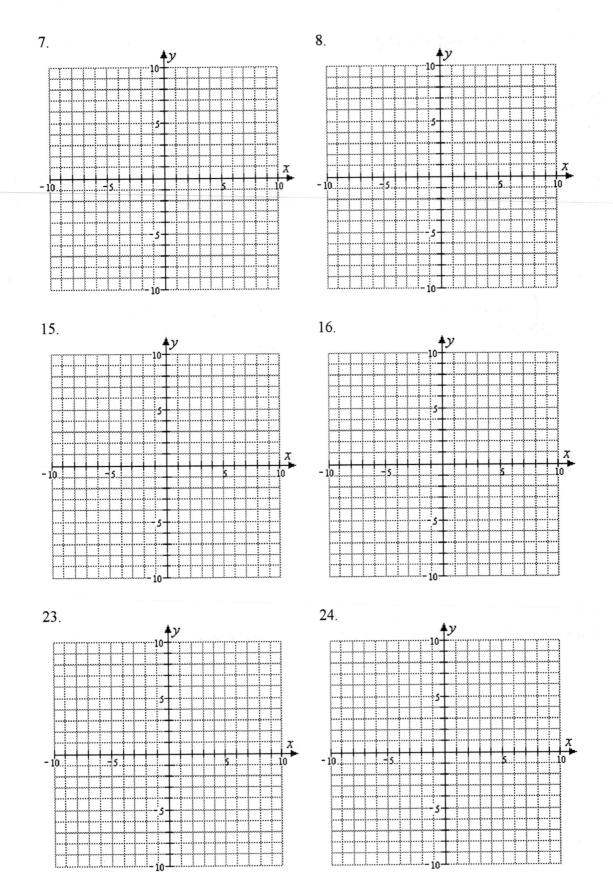

10.4 More About Modeling

Investigation 19: Periodic Functions

d	0	2	5	8	10	12	15	18	20	22	25	28	30	32	35	38	40	42	45	48	50
y																					

Table 10.13

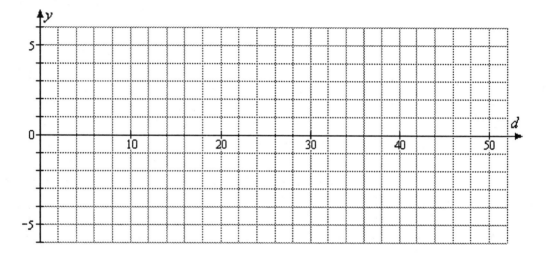

Figure 10.45

Exercise 1 Consider the swinging pendulum of a grandfather clock. Which of the graphs in Figure 10.0 best represents the height of the bottom of the pendulum as a function of time?

Exercise 2 Consider different rectangles in the first quadrant for which two of the sides lie on the coordinate axes, and the upper right vertex lies on the graph of $y = 12 - 2x$, as

295

shown in the figure. The shape of the rectangle depends on which point on the line $y = 12 - 2x$ is used as a vertex.

a. Find the area of the rectangle as a function of the x-coordinate of the upper right vertex.

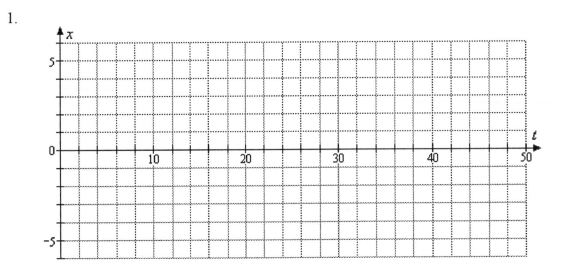

b. What is the domain of this function?

c. Graph the function on its domain.

d. What is the maximum value of this function?

Homework 10.4

1.

10.5 Joint Variation

Exercise 1   A retirement plan requires the employee to put aside a fixed amount each year until retirement. The amount accumulated (including 8% annual interest) depends on both the number of years contributing to the plan and the fixed contribution each year.

Number of years of contributions

		10	20	30	40	50
annual	500	7243	22,881	56,642	129,528	286,885
contribution	1000	14,487	45,762	113,283	259,057	573,770
	1500	21,730	68,643	169,925	388,585	860,655
	2000	28,973	91,524	226,566	518,113	1,147,540
	2500	43,460	137,286	339,850	777,170	1,721,310
	3000	50,703	160,167	396,491	906,698	2,008,196

a. How much will have accumulated after 40 years of contributing $500 each year?

b. How much must be put aside each year in order to accumulate $573,770 after 50 years?

c. How many years of contributions are needed to accumulate $137,286 making annual payments of $2500?

Exercise 2   Two stock market advisors are being compared based on their records from two years, 1990 and 2000. In each of those years, the two advisors recommended certain stocks, some of which turned out to be profitable investments and others that were not. The results are summarized in the table below.

Recommendations from Two Stock Market Advisors

	1990		2000	
	Advisor A	*Advisor B*	*Advisor A*	*Advisor B*
Profitable	32	36	45	14
Losers	28	28	23	6

a. How many total recommendations were made by advisor A? By advisor B?

b. How many of advisor A's total number of recommendations were profitable? How many of advisor B's recommendations?

c. What percentage of advisor A's recommendations were profitable? Advisor B's? Which advisor recommended the higher percentage of profitable stocks?

d. How many stocks did advisor A recommend in 1990? What percentage of these were profitable? What was Advisor B's record in 1990?

e. What percent of advisor A's year 2000 recommendations were profitable? Advisor B's?

Exercise 3 The cost of tiling a rectangular floor depends on the dimensions of the floor. The table shows the costs in dollars for some dimensions (length and width).

		Length (feet)					
		5	6	7	8	9	10
	5	400	480	560	640	720	800
Width	6	480	576	672	768	864	960
(feet)	7	560	672	784	896	1008	1120
	8	640	768	896	1024	1152	1280
	9	720	864	1008	1152	1296	1440
	10	800	960	1120	1280	1440	1600

a. How much does it cost for a floor that is 7 feet by 8 feet?

b. How much does it cost for a floor that is 5 feet by 9 feet?

c. Consider the row corresponding to 6 feet (in width). How does the cost depend on length?

d. Consider the column corresponding to (a length of) 10 feet. How does the cost depend on width?

e. The cost varies jointly with the length and width of the rectangular floor. Find a formula relating the base, height, and cost of tiling the floor.

Exercise 4 Use your formula from Exercise 3 to predict the cost of tiling a rectangular floor that is 12 feet wide and 15 feet long.

Homework 10.5

7.

	Velocity (m/s)			
Mass (kg)	1.0	2.0	3.0	4.0
1.0				
1.5				
2.0				
2.5				

8.

	Radius (ft)		
Height (ft)	20	25	30
30			
40			
50			
60			

9a.

	Pressure (atmospheres)							
Temperature (°C)	50	100	150	200	250	300	350	400
350								
400								
450								
500								
550								

c.

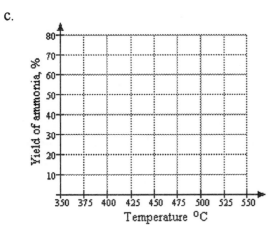

10a.

	Relative Humidity (%)					
Air temperature (°F)	0	20	40	60	80	100
80						
90						
100						
110						
120						

c.

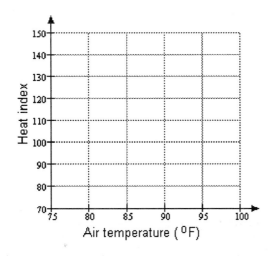

Chapter 10   Review

1.

2.

3.

4.

5.

6.

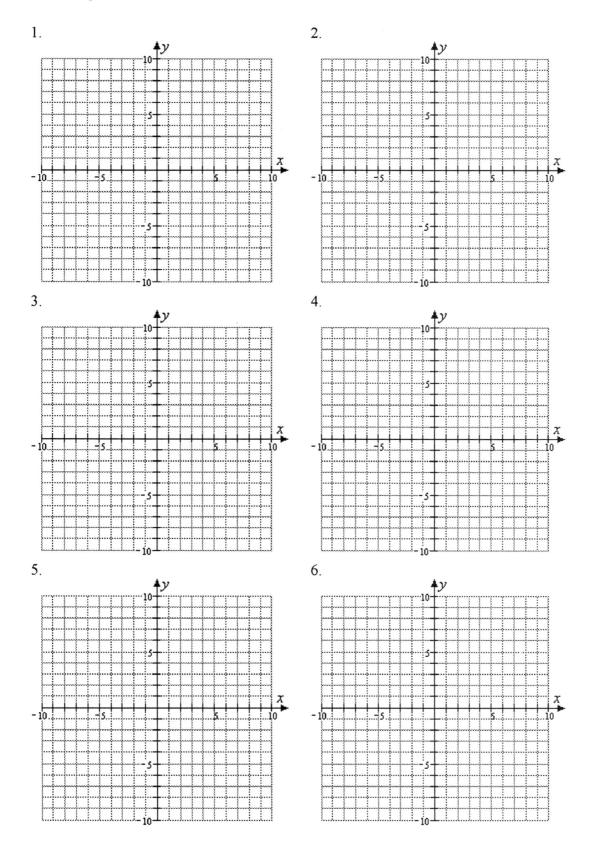

7.

8.

9.

10.

11.

12.

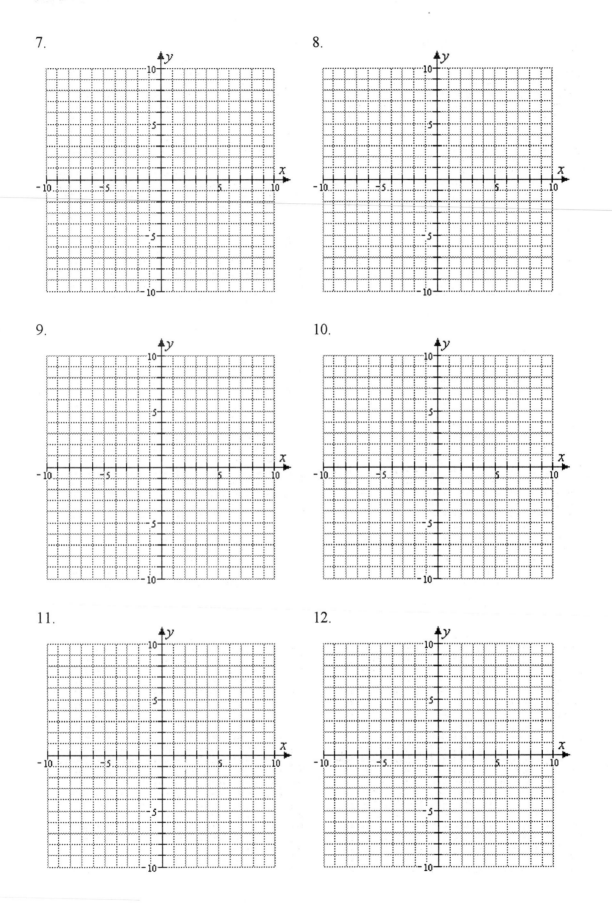

13.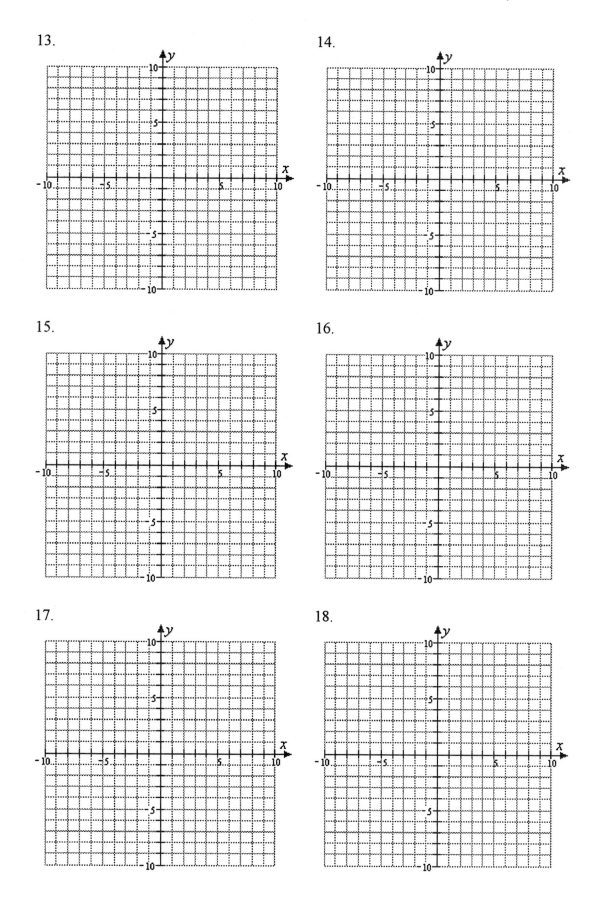

14.

15.

16.

17.

18.

25.

26.

27.

28.

29.

30.

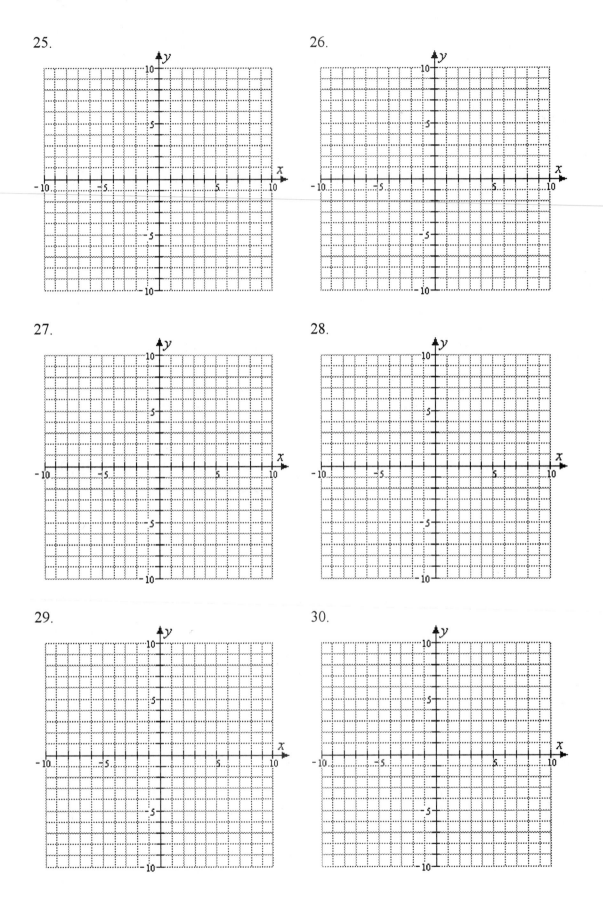

Chapter 11   More About Graphing

Investigation 20   Global Positioning System

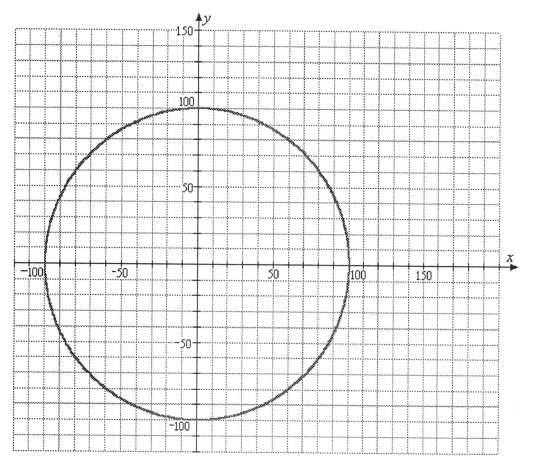

11.1 Conic Sections

Exercise 1   Graph $9x^2 + 8y^2 = 16$.

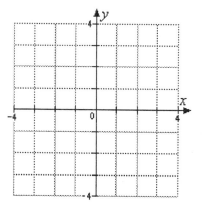

Exercise 2   Find the exact coordinates of all points with y-coordinate -1 on the ellipse $9x^2 + 8y^2 = 16$.

Exercise 3   Graph $\dfrac{x^2}{9} - \dfrac{y^2}{16} = 1$.

Find a and b, and plot the vertices.

Sketch the central rectangle.
Draw the asymptotes.
Sketch the hyperbola.

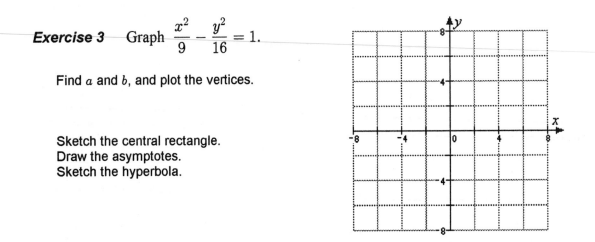

Exercise 4   Write the equation $4x^2 = y^2 + 25$ in standard form. Find the vertices of the graph, and the equations of the asymptotes.

Exercise 5   Solve the equation $4y^2 - x^2 = 16$ when $y = 1$. What is the significance of the solutions to the graph of the hyperbola?

Exercise 6 Graph $y^2 - 8x = 0$.

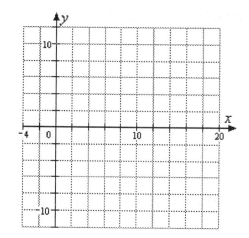

Homework 11.1

1.

2.

3.

4.

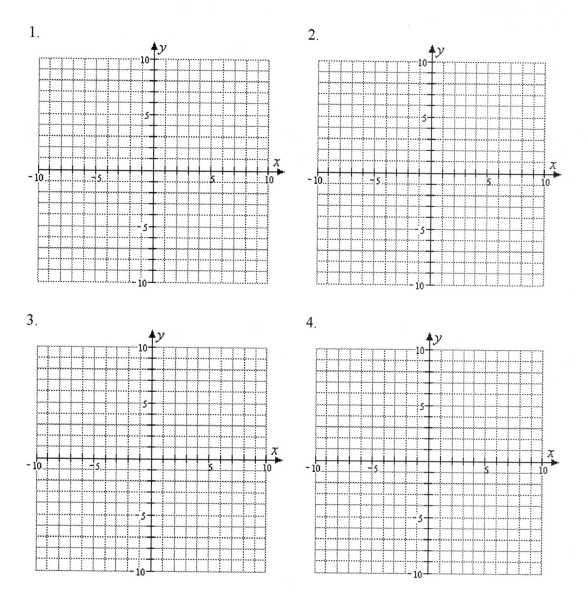

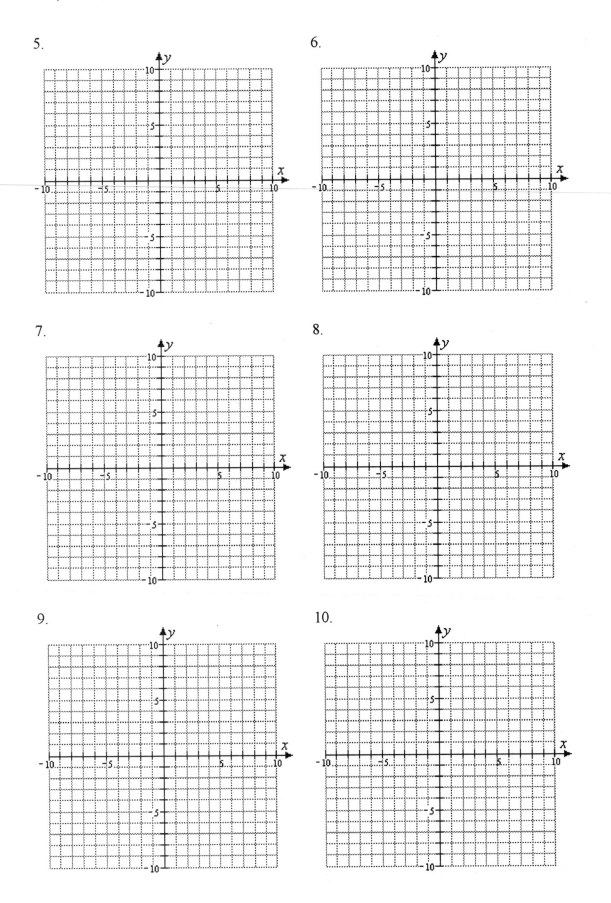

5.

6.

7.

8.

9.

10.

11.

12.

13.

14.

15.

16.

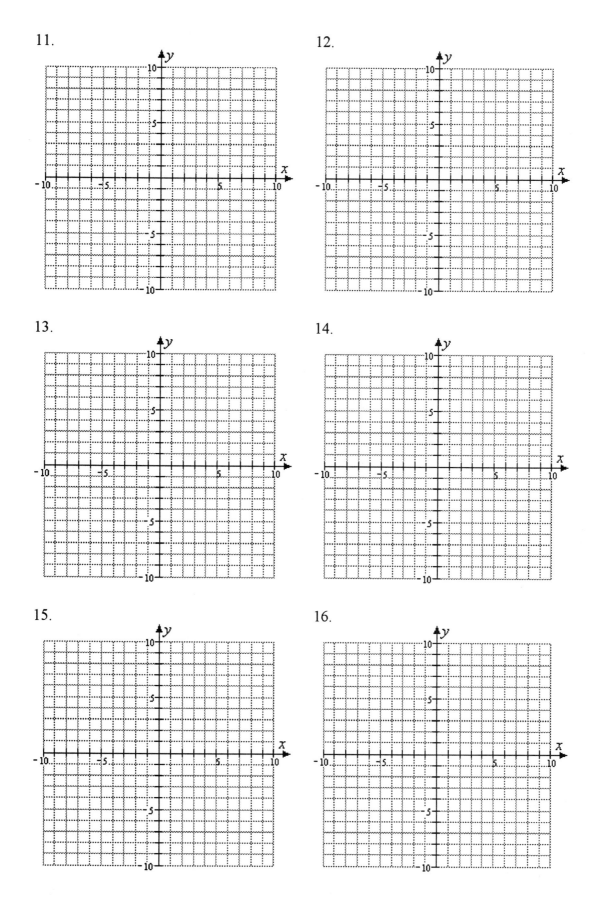

17.

18.

19.

20.

21.

22.

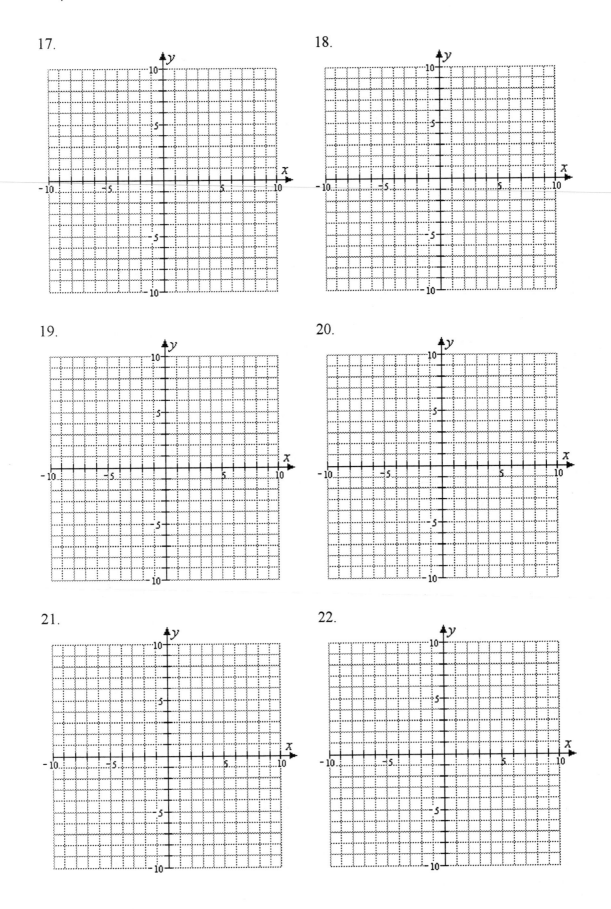

23.

24.

25.

26.

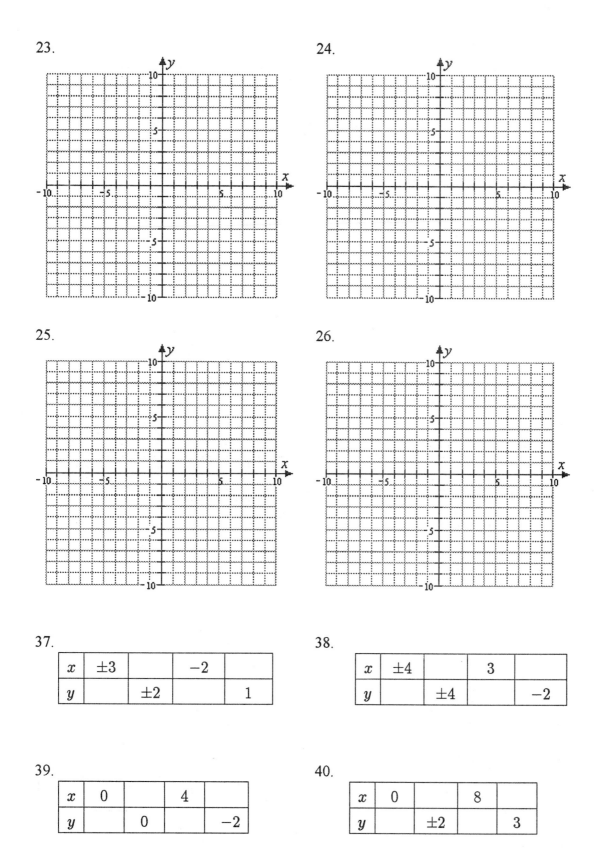

37.

x	± 3		-2	
y		± 2		1

38.

x	± 4		3	
y		± 4		-2

39.

x	0		4	
y		0		-2

40.

x	0		8	
y		± 2		3

53.

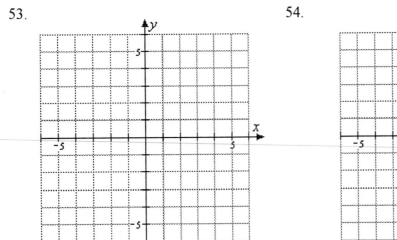

54.

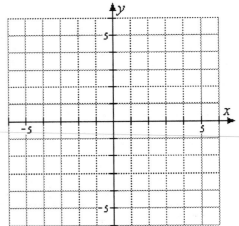

55.

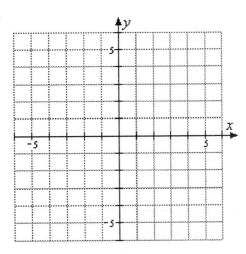

56.

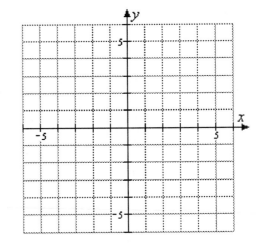

11.2 Translated Conics

Exercise 1a. Graph

$$\frac{(x-5)^2}{15} + \frac{(y+3)^2}{8} = 1$$

b. Find the coordinates of the vertices.

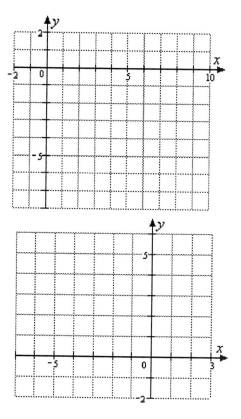

Exercise 2a. Write the equation

$$x^2 + 4y^2 + 4x - 16y + 4 = 0$$

in standard form.
b. Graph the equation.

Exercise 3 Find the equation of an ellipse with vertices $(-7, 1)$ and $(-1, 1)$ and a minor axis of length 4.

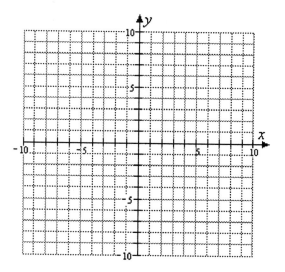

315

Exercise 4a. Graph $\dfrac{y^2}{9} - \dfrac{(x+4)^2}{12} = 1$.

    b. Find the equations of the asymptotes.

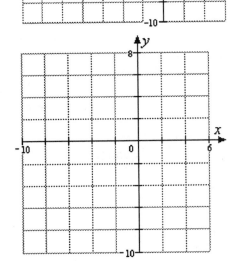

Exercise 5a. Write the equation

$$4x^2 - 6y^2 + 16x - 12y - 14 = 0$$

in standard form.

    b. Graph the equation.

Exercise 6a. Write

$$x = y^2 - 8y + 13$$

in standard form.

    b. Graph the equation.

Exercise 7   Describe the graph of each equation without graphing.

   a. $x^2 = 9y^2 - 9$

   b. $x^2 - y = 2x + 4$

   c. $x^2 + 9y^2 + 4x - 18y + 9 = 0$

Homework 11.2

1.

2.

3.

4.

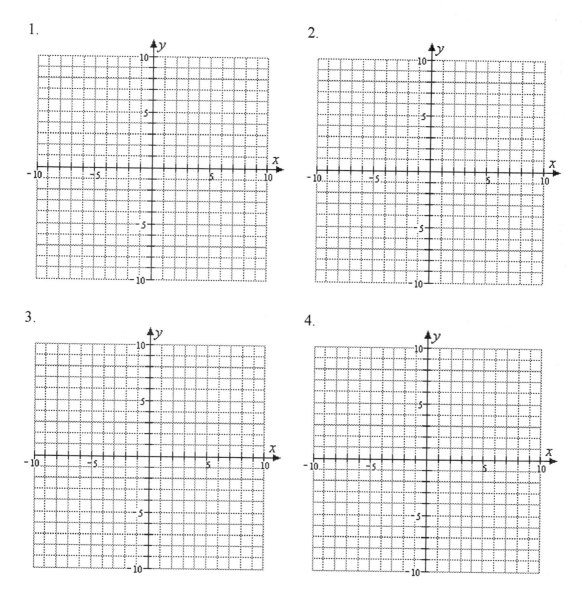

5.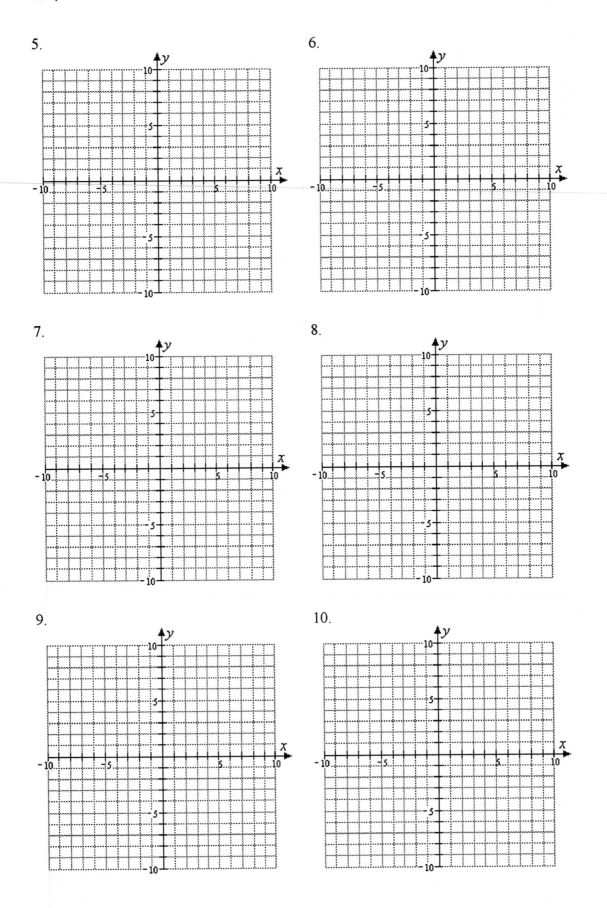

6.

7.

8.

9.

10.

17.

18.

19.

20.

21.

22.

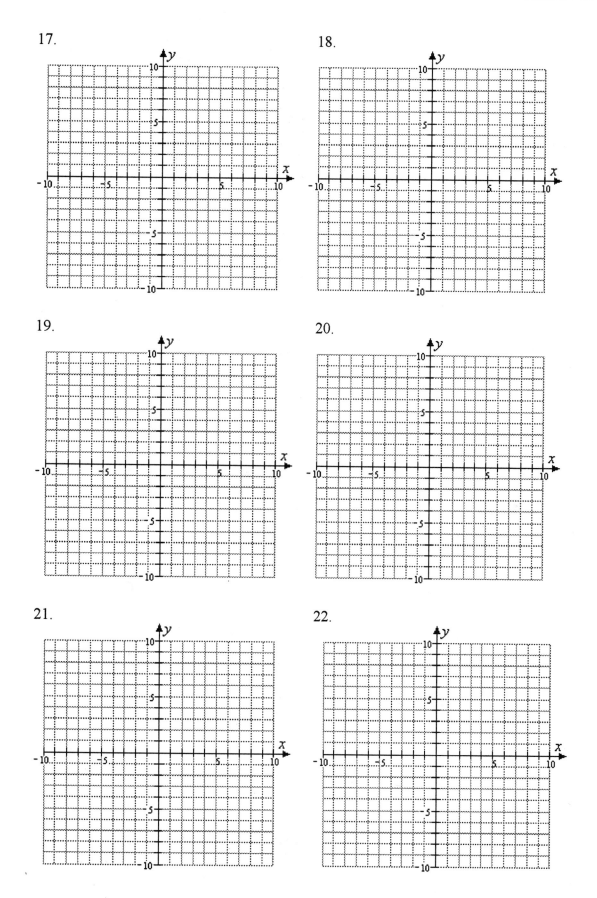

23.

24.

25.

26.

31.

32.

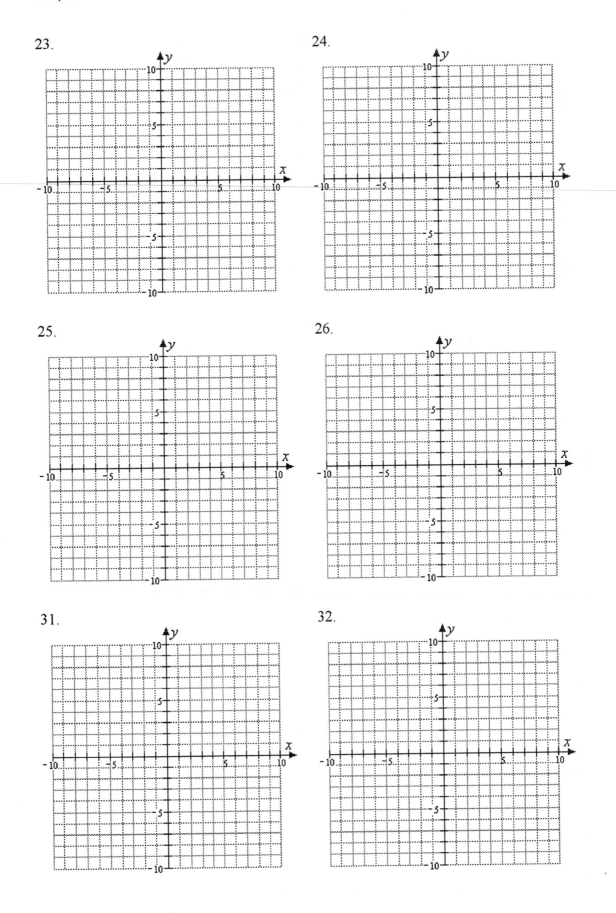

33.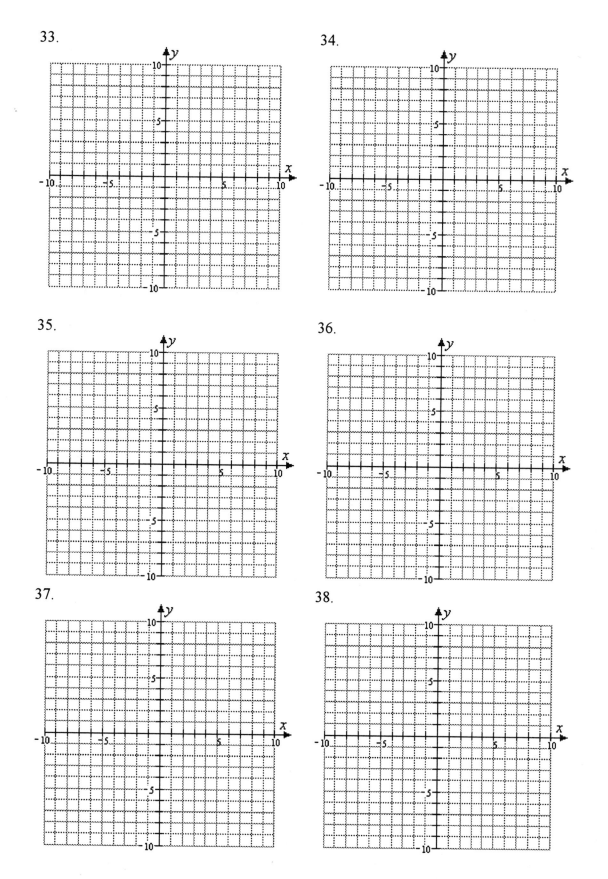

34.

35.

36.

37.

38.

11.3 Systems of Quadratic Equations

Exercise 1   Find the intersection points of the graphs of

$$x^2 - y^2 = 35$$
$$xy = 6$$

Exercise 2   Find the intersection points of the graphs of

$$y^2 - x^2 = 5$$
$$x^2 + y^2 = 13$$

Exercise 3 Find the intersection points of the graphs of

$$x^2 - 8x + y^2 + 2y = 23$$
$$x^2 - 12x + y^2 - 6y = -25$$

Homework 11.3

25.

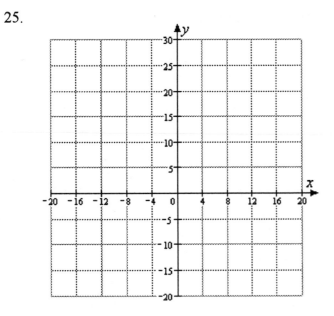

26.

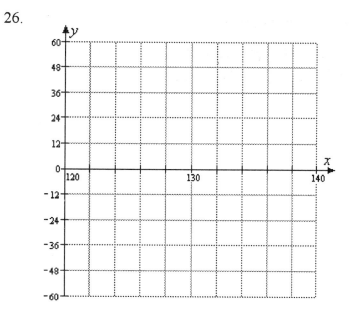

27.

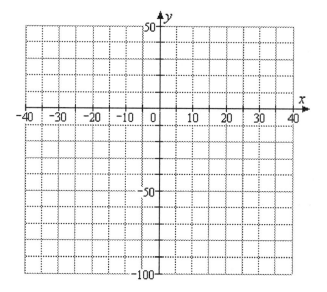

28.

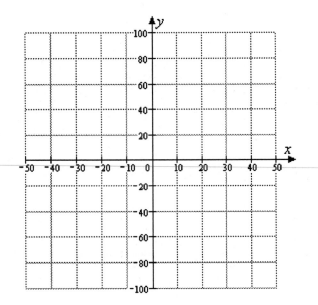

Midchapter Review

1.

2.

5.

6.

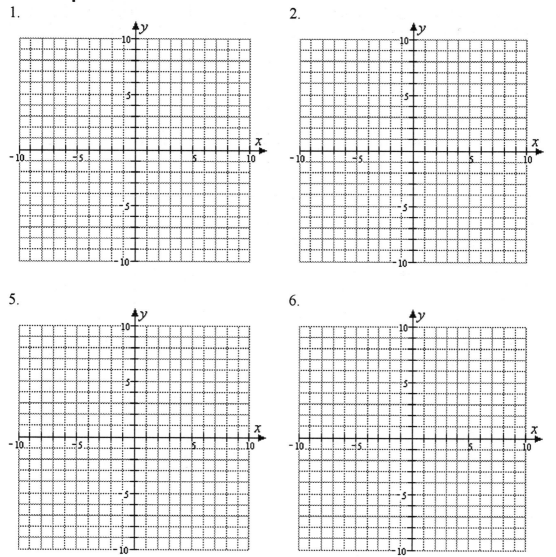

7.

8.

11.

12.

13.

14.

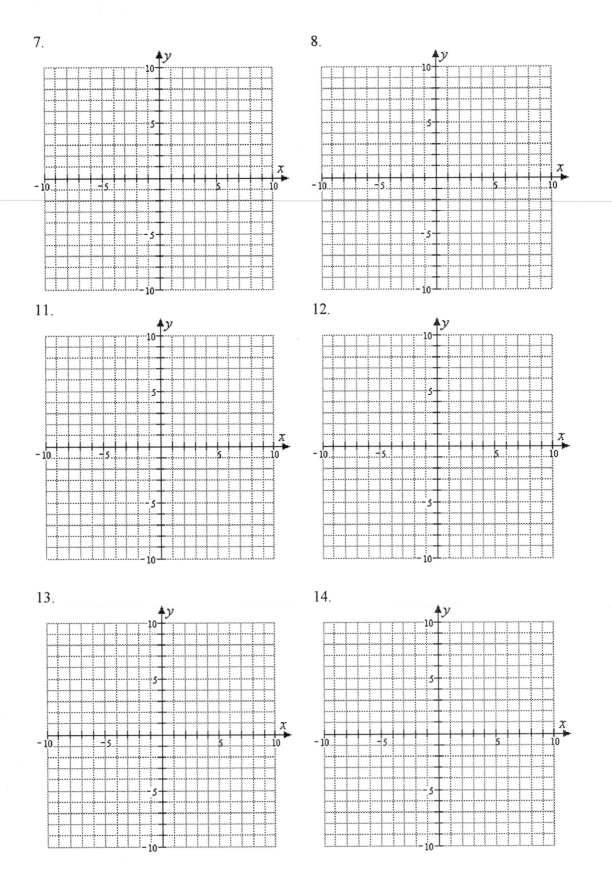

17.

18.

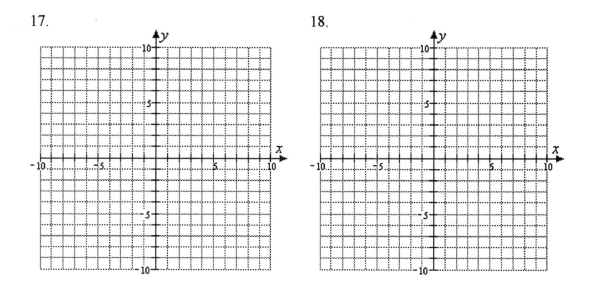

11.4 Linear Programming

Exercise 1 The figure shows the feasible solutions for a linear programming problem. The objective function is $R = x + 5y$.

a. Sketch lines for objective values of $R = 5$, $R = 15$, $R = 25$ and $R = 35$.

b. Evaluate the objective function at each vertex of the shaded region.

c. Which vertex corresponds to the maximum value of the objective function? What is the maximum value?

d. Which vertex corresponds to the minimum value of the objective function? What is the minimum value?

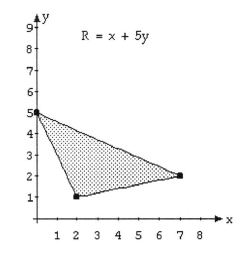

Homework 11.4

1-4.

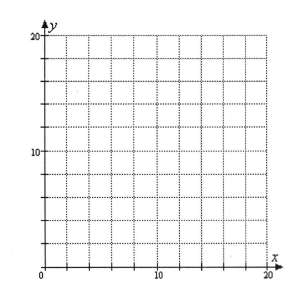

5-8.

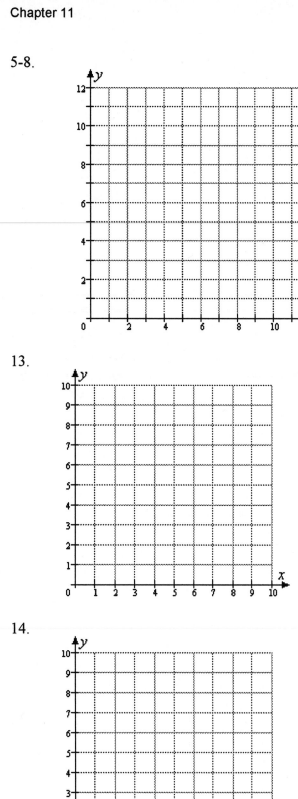

13.

14.

15.

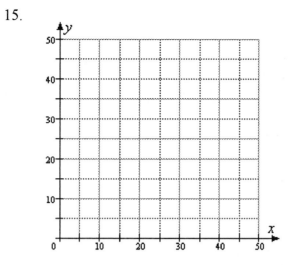

16.

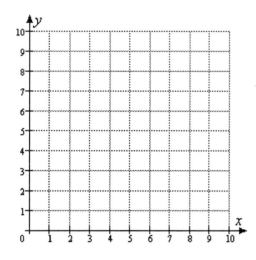

17.

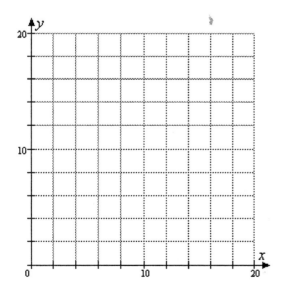

18.

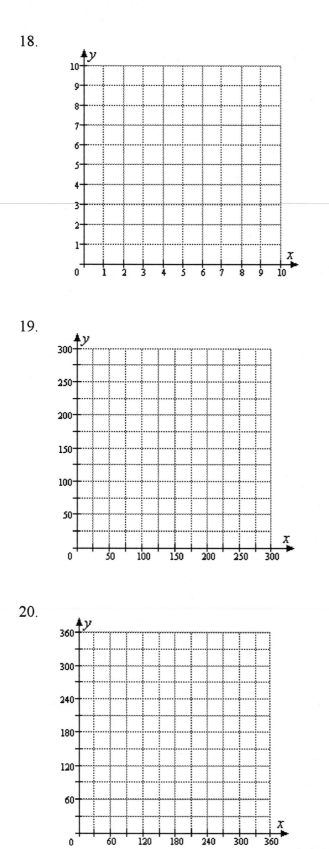

19.

20.

21.

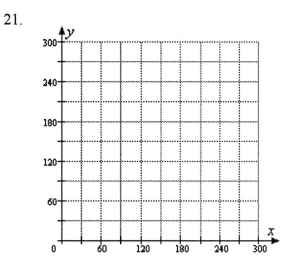

22.

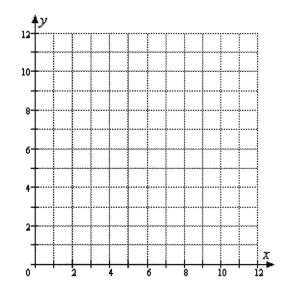

23.

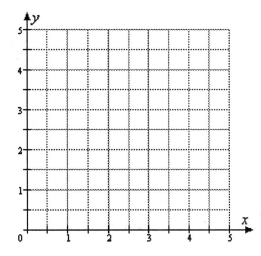

24.

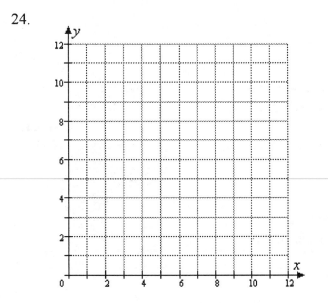

25.

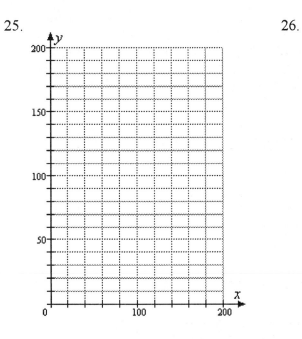

26.

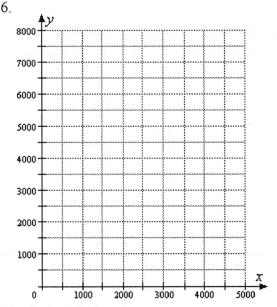

11.5 Displaying Data

Exercise 1   The pie chart shows the market share of major recording companies for sales of music CD's and cassettes.

   a. What is the ratio of Universal's market share to EMD's share?

   b. What percent of the market belongs to "Other" companies? How large is the angle for the sector representing "Other"?

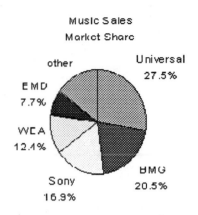

Exercise 2   The histogram shows predicted EPS (earnings-per-share), in dollars, for companies with the greatest annual growth.

   a. How many companies are depicted in the histogram?

   b. Find the mode, mean, and median of EPS for these companies.

   c. What is the relative frequency of an EPS of $0.65?

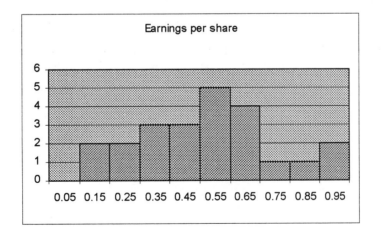

Exercise 3   The boxplot depicts the spread of earnings-per-share (EPS) among a sample of 50 companies.

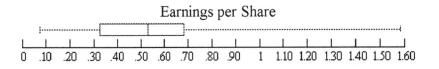

Earnings per Share

a. Approximately what percentage of the companies have an EPS exceeding $0.32?

b. If these companies are representative of companies on a stock exchange, what fraction of the companies on the exchange have an EPS between $0.32 and $0.68?

Exercise 4   The contents of pretzel packages have normally distributed weights. The mean weight is 12 ounces and the standard deviation is 0.05 ounces.

a. What fraction of the pretzel packages contain between 11.9 ounces and 12.1 ounces?

b. What fraction of the packages contain more than 12.15 ounces?

Chapter 11   Review

1.

2.

3.

4.

5.

6.

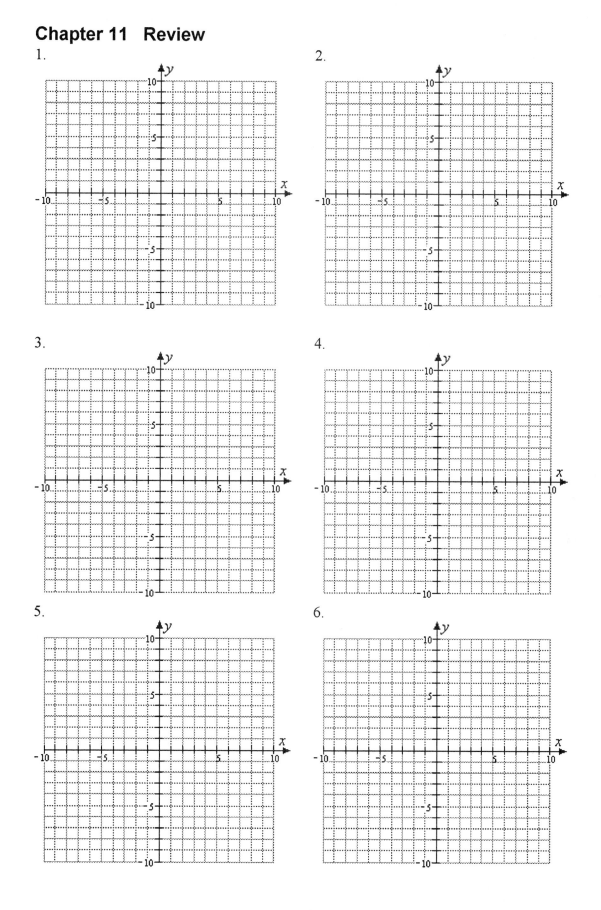

7.

8.

9.

10.

33.

34.

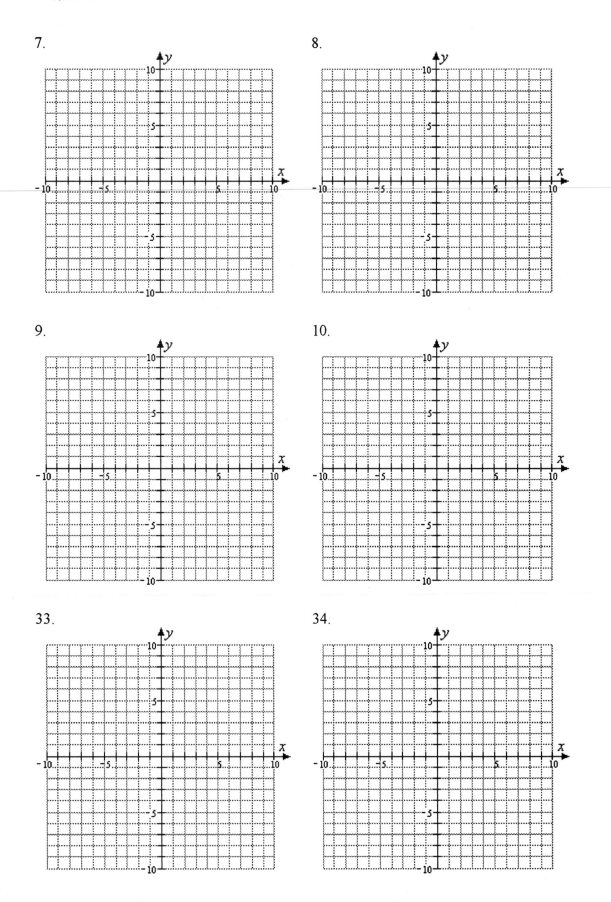

35.

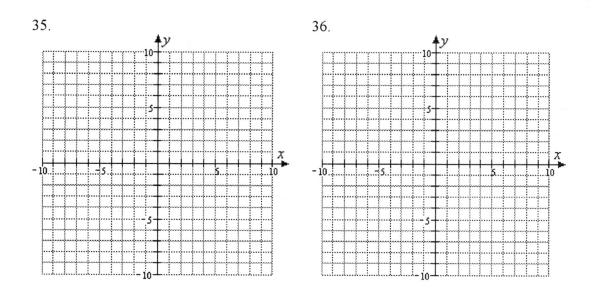

36.

43.

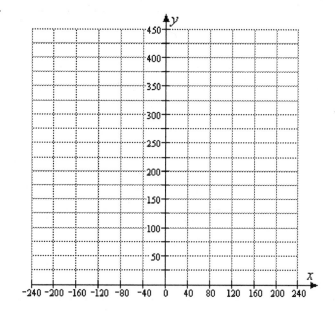

44.

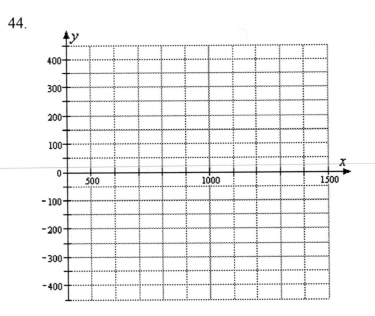

45.

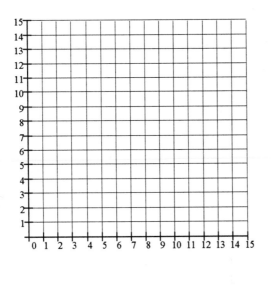

46.

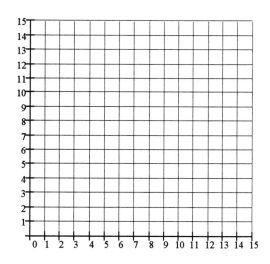

47.

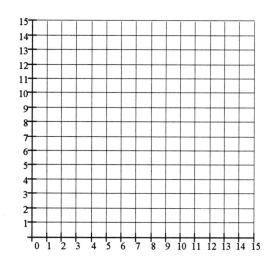

48.

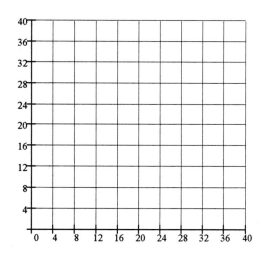